U0271113

符号中国 SIGNS OF CHINA

中国木文化

CHINESE WOOD CULTURE

“符号中国”编写组 ◎ 编著

中央民族大学出版社
China Minzu University Press

图书在版编目(CIP)数据

中国木文化：汉文、英文 / “符号中国”编写组编著. — 北京：中央民族大学出版社, 2024.3
（符号中国）
ISBN 978-7-5660-2279-0

Ⅰ. ①中…　Ⅱ. ①符…　Ⅲ. ①木材–文化–介绍–中国–汉、英　Ⅳ. ①S781

中国国家版本馆CIP数据核字（2024）第016516号

符号中国：中国木文化 CHINESE WOOD CULTURE

编　　著　“符号中国”编写组
策划编辑　沙　平
责任编辑　杨爱新
英文指导　李瑞清
英文编辑　邱　械
美术编辑　曹　娜　郑亚超　洪　涛
出版发行　中央民族大学出版社
北京市海淀区中关村南大街27号　邮编：100081
电话：（010）68472815（发行部）　传真：（010）68933757（发行部）
（010）68932218（总编室）　（010）68932447（办公室）
经 销 者　全国各地新华书店
开　　本　787 mm×1092 mm　1/16　印张：10.5
字　　数　136千字
版　　次　2024年3月第1版　2024年3月第1次印刷
书　　号　ISBN 978-7-5660-2279-0
定　　价　58.00元

本册编写者

尚　景

从古至今，木材都是中国人生活中不可或缺的原料，从木建筑、木家具，到各种木制用品，木与人们的生活息息相关、如影随形。中国人之所以对木情有独钟，主要是因为木材取自自然，其颜色、纹理、质地均富有天然的美感与质感。此外，木材软硬适中，易于加工和雕刻。

中国人不仅用木、爱木，还赋予木制品深厚的文化意味。对木材的打磨加工和精雕细刻，记录了数千年的中华文化与历史。一件件巧夺天工

Since ancient times, wood has been an indispensable raw material in the life of Chinese people. From wooden buildings and wooden furniture to various other kinds of woodenware, wood has been closely linked with Chinese people's daily life. Chinese people's love of wood mainly stems from its natural origin, which gives natural beauty and quality to its color, grain, and texture. In addition, wood is neither too soft nor too hard and is therefore easy to process and carve.

的木制品、木家具以及或雄伟或精致的木建筑，承载着中国人的智慧和中华数千年的文化底蕴。

本书用精美的图文，将中国的木建筑、木家具、木雕艺术与文化展现出来，相信一定能让您了解木，了解中国的木文化，更通过它来了解中国。

The Chinese people not only use wood, love wood, but also give woodwork profound cultural charm. Finely ground and precisely carved, the wooden artwork records every detail of the thousand-year-old Chinese culture and history. In form of tasteful wooden furniture, majestic or exquisite wooden buildings, or woodwork of various kinds, they bear witness to the wisdom of Chinese people as well as thetime-honored culture of the Chinese nation.

With elegant pictures and essays, this book introduces the wooden buildings, wooden furniture, as well as art and culture of wood carving of China. It will surely bring the readers one step closer to the understanding of wood, wood culture of China, and even China as a whole.

目 录 Contents

源远流长的木文化
Long-standing Wood Culture

中国人自古以来就喜欢木，崇尚木，几千年来对它始终情有独钟。在古代中国人看来，木是生命之源，发自春天，发于自然。中国人追求“天人合一”，而“木”恰恰源于天然，因而被运用得颇为精妙。宏伟壮阔、简练淳朴、巧夺天工、玲珑雅致……“木”被广泛应用于建筑、家具、生活用品等各个方面，与人们的生活息息相关，木文化也因此蓬勃发展起来。

Chinese people have been loving and adoring wood since the ancient times. Such affection has lasted for several millenniums. For the ancient Chinese, wood is the source of life. It originates in spring from Mother Nature, fitting in well with Chinese people's pursuit of the unity of man and nature. Therefore, they have used wood to the exquisite limit and created countless works of art that are magnificent and grandiose, simple and pure, and skilful and dainty … Wood has been used to make buildings, furniture, and articles of everyday use. It is a material closely linked with people's daily life. Consequently, wood culture has been thriving.

> 中国悠久的木文化

古代中国人种树为林，伐木取材，运用高超的工艺将取自天然的“木”盖成房屋、制作成家具、雕刻成器物。吃、穿、住、用、行，每个生活细节无不与“木”息息相

• 关于原始人生活的油画

原始人将木材制成木棒，用来捕猎野兽、获取食物或保护同伴的安全。

An oil painting about the life of the primitives

The primitives made stick from wood to catch wild animals for food or to protect their fellows.

> Time-honored Wood Culture of China

The ancient Chinese planted trees for wood. Their marvelous skills turned wood into houses, furniture, and carved utensils. Wood has infiltrated into every aspect of people's daily life: eating, dressing, living, tool using, and traveling. It is safe to say that Chinese people have a millennium relationship with wood.

From Fire Making by Drilling Wood to Tree Felling for House Building

The primitive men used wood first as fuel after their discovery of the dead animals killed by forest fire. Since then, they began to eat cooked food. Later, they began to drill a hard piece of wood against another to get the spark for starting a fire. Then, they found that a fallen tree could support their thatched huts, giving them better shelters against

关。可以说，中国人与“木”有着千年之缘。

从钻木取火到伐木造屋

远古时期的人类最初开始用木，是在森林起火后，发现大量的野兽被烧死，原始人开始吃熟的食物。后来，人们试着将坚硬的木头在另一块木头上用劲钻，钻出的火星可以点燃茅草。再后来，人类又发现倒下的大树可以支撑起茅草屋，以此来遮风挡雨、驱寒保暖。逐渐地，人类开始越来越多地使用木，并将其作为建造房屋的首选良材。人类文明也正是从这最初的草木房屋衍生开来。

“木”字起源

中国古老的文字甲骨文（甲骨文：中国已发现的时代最早、体系较为完整的古代文字，是殷商时代刻在龟甲兽骨上的文字）是象形文字，而甲骨文中最早的“木”字，形体是树的枝丫，上面是枝叶，下面是树根，十分形象，后来渐渐演变成今天的“木”字。

中国人自古以来就崇尚线条

wind, rain, and coldness. Gradually, they put wood to more and more uses. However, house building remained the prime usage of wood. In fact, human civilization started just from the most primitive thatched houses.

Origin of the Chinese Character 木 (Wood)

The ancient Chinese character Oracle Bone Script (*Jiaguwen*), is a pictograph. (*Jiaguwen*: The earliest Chinese character in a relatively complete system, are the inscriptions on bones or tortoise shells appeared in the Shang Dynasty (1600

原始人的草木房屋
A Thatched Hut of the Primitives

美，尤其是书法，将文字的功能与审美巧妙结合起来。而“木”字的一横一竖，一撇一捺都是线条，就像是一座房屋的雏形，通过柱和梁展现一个“木”的世界。

独树成“木”，双木成“林”，三木成“森”，五木即成“森林”。古人从森林中伐木，使之变成材，而后造成房屋、家具及生活用品。所以现在的很多汉字，尤其是木制的物品，其字皆有“木”旁。

木与阴阳五行

阴阳五行是中国古代最具影响力的学说之一。古人认为，大自然主要由五种元素构成，即金、木、

B.C.-1046 B.C.). The character 木 (wood) was among the earliest Jiaguwen words. It has a vivid tree-like shape, with the upper part representing the branches and the lower part as the roots.

Chinese people have always loved the beauty of the lines. Chinese calligraphy, in particular, combines the communication function and aesthetic value of characters. The character 木 (wood) contains a horizontal, a vertical, a left-falling stroke, and a right-falling stroke, all being lines. They form a rouge shape of a house as together. Through the column and beam, we can see a world of wood.

Putting two characters 木 (wood) together side by side makes a new character 林 , which means woods.

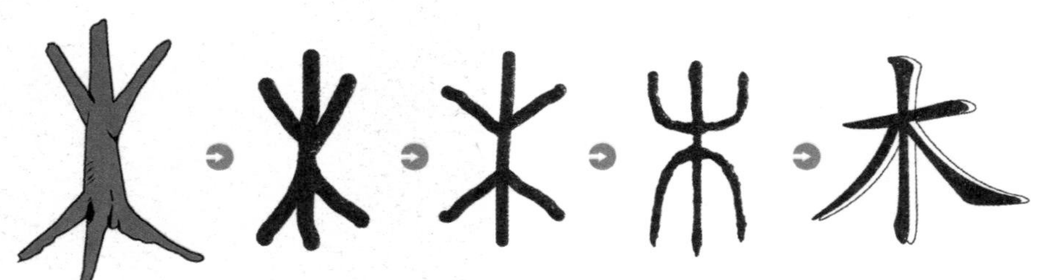

• **“木”字的发展**
Development of the Character 木 (Wood)

• **北京故宫中和殿**

中国宫殿倾斜的屋顶好像一撇一捺，线条美十足。

The Hall of Central Harmony in the Imperial Palace (the Forbidden City), Beijing

The tilting roofs are like a left-falling stroke and a right-falling stroke, showing the complete beauty of the lines.

水、火、土。木主春天，具有生发性，代表着生生不息。

中国人的家庭观念很重，人们把“家”视为休养生息的立命之所，而社会则是由千万个“小家”组成的“大家”。“家”的用材、布局、环境等方面都离不开“木”。

Interesting enough, piling three 木 (wood) up will get the character 森, which is safe to guess it must be forest. Therefore, you can see Chinese word 森林 (forest) is made up of nothing but five 木 (wood) The ancient Chinese felled trees from the forest and used the wood to build houses and make furniture and articles of everyday use. Many modern Chinese

• 苏州惠和堂内的古车
An Ancient Cart in Huihe Mansion, Suizhou

伐木

早在原始社会，人类就已经开始伐木取材建造房屋。

古代中国人对伐木是很讲究的。起初，古人只在冬天伐木，因为这样不耽误农时，而且冬天虫蚁滋扰较少，便于砍伐、运输。后

characters, especially those for the articles made of wood, have 木 (wood) as a component.

Wood and Yin-Yang and Five Elements

The theory of Yin-Yang and Five Elements is the most influential one in ancient China. The ancient people believed that the nature was composed of five elements, namely Metal, Wood, Water, Fire, and Earth. Wood also indicated spring, boosted growth, and represented the thriving force of life.

Chinese people have had an extremely firm attachment to their families. They treated home as a place for rest and recuperation and the society as a big family composed of millions of tens of small families. The ancient people regarded wood as the foundation of their home, because none of the elements of their home, such as its building, layout, and environment, could do without wood.

Tree Felling

As far back as the primitive society, mankind began felling trees to get wood for house building.

For ancient Chinese people, tree felling was done in a well-planned way.

来，随着对树木生长规律的深入了解，人们开始根据树种的不同来分季节采伐，一般的木材适宜于四月或七月采伐。

古人在伐木时，要先确定树木倒下去的方向，防止其损伤种树和幼树，也是为了便于运输，同时保证伐木人的安全。人们先将周围障碍物清理干净，规划出一条安全通道，再将树伐倒。伐倒的树木，先要去掉枝丫，然后按不同长度锯成“造材”，即原木。最后，再利用冰雪天气、河道、滑道、牲畜、车马或人力等方式将原木运出山林。

• 木匠正在运用工具打造木家具
A Carpenter is Making Wooden Furniture with a Plane.

At first, the ancient people felled trees only in winter when they had few farm works. Moreover, it was covenient for people to fell trees and carry woods due to less harassment from bugs. Then, with their knowledge about trees accumulating, people began felling different types of trees in different seasons. Generally speaking, ordinary trees are good to be felled in April or July.

When felling trees, the first thing for the ancient people was to choose a direction the tree will fall down. They did this to avoid damaging seedlings and young trees in the vicinity, and to guarantee lumberers'safety. And it would be easier to move big logs. Then, they cleared away all the obstacles around and cut out a safe transport path. After the tree was felled down, they cut the branches and twigs off the trunk and sawed it into different lengths of bucking, also known as log. Finally, they transported the logs out of forest by using icy path, river, chute, draught animals, carts, or labour power depending on the weather or circumstance conditions.

Carpenters made beams, columns, rafters, doors, and windows out of logs and then built up houses. More than that, they also made furniture, musical

工匠们把一根根原木制作成梁、枋、椽、门、窗，最终盖起一座房屋，还将木材做成家具、乐器、车船等，构建出中国辉煌的木建世界。木材已不仅仅是树产品，更是人类的“朋友”。

instruments, carts and boats. Their hard work created a glorious world of Chinese woodwork. With such contribution, wood is not only a product of trees, but a friend to human.

鲁班

鲁班是中国木工的“祖师爷”，春秋时鲁国人。有一次爬山的时候，鲁班的手指不小心被一棵小草划破了。他很奇怪：为什么小草如此锋利？于是仔细察看，发现草叶两边全是排列均匀的小齿。后来，他就模仿草叶制成了锯，中国的木文化也由此发扬光大起来。鲁班在建筑、木工、器械等方面有大量发明创造，如建筑用的曲尺、刨、钻、铲等。此外，鲁班还擅长雕刻。

Lu Ban

Lu Ban, a skilful craftsman in the state of Lu in the Spring and Autumn Period (770 B.C.-476 B.C.), was the founder of Chinese carpentry. Once, during a trip up the mountain, his finger was cut open by a blade of grass. He wondered why the grass blade was so sharp. A close look revealed that the blade was lined on both sides with even serrations. Later, Lu Ban invented saw by imitating the serrations on the grass blade. Since then, Chinese wood culture has been greatly developed and widely spreaded. Lu Ban also made many other inventions in architecture, carpentry, and instrument making, including the building tools such as bevel gauge, plane, drill, and shovel. In addition, Lu Ban was good at carving as well.

> 木的特质与分类

木源于树，主要取自树干，而树干又由树皮、形成层、木质部、髓心组成。木材之所以被人们广泛应用，是因为它具有一些其他材料无法比拟的特性。

取自天然

中国人自古以来就遵循天人和谐之道，崇尚自然，而木恰恰生长于自然。木材的生产成本较低、耗能小、无毒害、无污染，因而被广泛应用于建筑以及日常生活中。中国古代，人们在使用木材时，更多地考虑怎样与自然相和谐。建筑布局中的依山就势、见水修桥、因高建阁、依地为宫等，均是因地制宜，依赖于自然山水，形成了木建筑和木器具的独特风格。

> Characteristics and Classification of Wood

Wood originates from the tree, especially the tree trunk. Tree trunk consists of bark, cambium, xylem layer, and pith. Wood has been widely used by people because of its many unparalleled features.

From Nature

Chinese people have been following the way of harmony and cherishing

• 木质佛珠
A Wooden Buddhist Rosary

树皮：树干的最外层。

Bark: It is the outer layer of the tree trunk.

形成层：位于树皮与木质部之间，包裹整个树干，形成层的木材不结实。

Cambium: Between bark and xylem, cambium covers the entire tree trunk. The wood of cambium is not robust.

• **树的构造**

Structure of Tree Trunk

木质部：是树干的主要部分，也是木材的主体。靠近内层的木质层是成熟的木材；靠近外层的木质层是木质化程度较差的木材。

Xylem layer: It is the main part of the tree trunk and also the main body of wood. The xylem layer close to the inner layer is mature wood while that close to the outer layer is the poorly-lignified wood.

髓心：俗称"树心"，位于树干中央或偏离中央；颜色较木质层或深或浅些，材质松软，易开裂或腐朽。

Pith: It is in the center of the tree trunk or somewhere nearby. The color is deeper or lighter than xylem layer. It has a loose and soft texture and is apt to crack or rot.

北京北海公园里的古树
An Ancient Tree in Beihai Park of Beijing

Mother Nature since the ancient times. They naturally fall in love with wood, a product from nature. Moreover, wood production involves low cost and small energy consumption, and causes no poison and no pollution, making it a common material for building and daily use. In ancient China, people usually paid greater attention to staying in harmony with nature when using wood. They built structures by following the terrain and keeping them in line with local conditions: bridges over water, pavilions on high grounds, and palaces on right positions. The architectural layout not only depended on natural landform, but

装饰性强

木材有非常好的触觉特性和视觉特性，有天然美丽的花纹，色调柔和。同时，作为家具和装饰材料，木材还具有很好的装饰性，因为其软硬适中，易于加工。此外，

木雕艺术品
A Piece of Wood Carving Artwork

清代红木雕狮纹半桌和凳子
A Lion-grain Half Table and Two Stools Carved on Mahogany in the Qing Dynasty (1616-1911)

组成木材的管胞、导管、木纤维等细胞都有细胞腔，这使木材具有多孔性特点。木材的多孔性使其有很大的优势：导热性低；具有弹性，木材受重载荷冲击时能吸收相当部分的能量；容易锯解和刨切；有一定的浮力。

木材的缺点

当然，木材也有一些缺点。比如容易干缩湿胀和变形开裂；易受木腐菌、昆虫或海生钻木动物的危害而变色、腐朽或蛀蚀；易燃；干燥缓慢，易发生开裂、翘曲、表面硬化、溃烂等；不能像金属那样容易按人们意愿制成宽大的板材等。

also demonstrated the unique style of wooden buildings and wooden utensils.

A Good Decorative Material

With beautiful natural patterns and a gentle hue, raw wood has a very good touch and appearance. Made into furniture, it performs well as a decorative material. Its moderate hardness allows an easy processing. In addition, wood cells such as tracheid and vessel, and wood fiber all have cell cavities, which grant wood many favorable features of a porous material, including low heat conductivity, good elasticity, and certain buoyancy. Moreover, wood can absorb considerable proportions of the energy under the impact of a heavy load. It is easy to saw and cut.

什么是树木的年轮和生长轮？

树木在生长过程中，由于气候的交替变化，在树干内部形成了一种轮状结构。即树木在一个生长周期内，形成层向内分生的一层次生木质部，它围绕着髓心构成同心圆，称为"生长轮"。

年轮也是一种生长轮。年轮的宽窄随树种、树龄和生长条件而异。如椿木的年轮很宽，而黄杨木、紫檀木的年轮一般很窄。年轮的宽窄、明显度、形态是识别木材的主要方法之一。

What are tree annual ring and growth ring?

When a tree grows, a ring-like structure takes shape inside the tree trunk in response to climate alteration and change. Such structure is a series of concentric circles around pith. It forms after the growth of a layer of secondary xylem from the cambium within a growth cycle of the tree. It is called the growth ring.

The annual ring is a kind of growth ring. The width of the annual ring varies with tree species, tree ages, and growth conditions. Ailanthus, for example, has a very wide annual ring while boxwood and red sandalwood only have narrow annual rings. The width, evidence, and shape of the annual ring are among the main methods of identifying wood.

木材的分类

中国古典木工行业中，木材被分为两类，即硬木和软木。一般认为，针叶材材质较软，就是软木；阔叶材材质较硬，就是硬木。其实这一说法并不准确，因为有些针叶材如落叶松等，材质是坚硬的。中国木工行业的木材分类是约定俗成的。一般来说，硬木的密度较高，纹理较细，质量也较重，因而比软木更为名贵，其中的紫檀木、黄花

Drawbacks of Wood

Of course, wood also has some drawbacks. It shrinks when dried and swells when getting wet and is easy to deform and crack. It is susceptible to the harms by domestomycetes, insects, or marine borers, which result in color change, decay, or moth damage. It is combustible, slow in drying up, and easy to show crack, warp, surface hardening, and rot. It cannot be made into large board at will as easy as metal.

梨木等价格十分昂贵，是高档木材。而软木一般质地松软，伸缩性强，材质耐久稳定，适合制作小型家具与细部的雕琢，如有装饰花纹的雕刻等，因此在使用时多和硬木搭配。软木的种类繁多，一般价格比较低。

• 紫檀木切面
A Section of Red Sandalwood

Classification of Wood

In China's classical carpentry industry, wood has been divided into two types: hardwood and softwood. People generally believe that coniferous wood is soft and should be softwood, otherwise, broad-leaved wood is hard and should be a hardwood. However, such opinion is incorrect. Some coniferous wood such as larch is hard. In China's carpentry industry, wood classification is accepted through common practice. Generally speaking, hardwood has high density, a fine texture, and heavy weight, hence its higher value than softwood. Of hardwood, red sandalwood and yellow rosewood are top-grade wood and they are very expensive. Softwood is normally soft and has small elasticity. It is stable and lasts long and is therefore suitable for making small furniture and detailed carving, like that of decorative patterns. In practice, it is often used together with hardwood. There are many kinds of softwood and they are normally cheap.

> 中国人对木的应用

用于建筑

中国古代建筑的主要用材就是木，且木建筑的种类很多，有宫殿、园林、寺院、道观、桥梁、塔刹等，其架构大都以木结构为基础。比如一座宫殿，需要用到数万件木构件，但却一根钉子也不用就能紧密地搭建连接在一起，历经数百年甚至数千年风雨仍然屹立不倒。几乎适用于任何工艺的木结构，使中国建筑具有丰富的造型。彩画、镏金、琉璃、雕刻等装饰工艺与之互相配合，让中国的木建筑更具有独特的风格。

门和窗在中国古代通常是全由木材制成的建筑构件。古代的中国人非常重视大门，因为它是整座宅

> Use of Wood by Chinese People

For Building

The ancient Chinese buildings were mainly built with wood and they were in various types, including palace, garden, temple, Taoist abbey, bridge, and pagoda. Most of them had a wooden structure. A palace, for example, was composed of tens of thousands of wooden components and they were tightly joined together without the need of a single nail. Some of these buildings are still firmly standing today after several centuries or even several millenniums. That almost all building techniques are applicable to wooden structure has contributed to the rich shapes of Chinese buildings. Wooden decorative techniques such as color painting, gold plating, glazing, and carving are frequently used in wooden

顶部中间是雷公柱。

The Suspended Column stands in the middle and on top of the building.

雷公柱四周的4根通天木柱通体红色，绘有金色纹饰。

The four towering wooden columns around the Suspended Column are red from bottom up and bear golden decorative patterns.

殿内还有28根楠木大柱，支撑着整座大殿的重量。

Inside the hall are also 28 Phoebe zhennan columns to support the weight of the entire hall.

殿外不设墙体，用红色木质门窗代替。

The hall has no walls, but red wooden doors and windows instead.

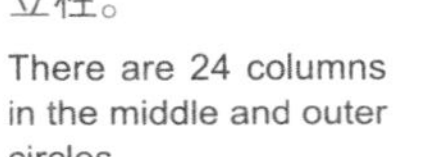

中围、外围有24根立柱。

There are 24 columns in the middle and outer circles.

- **北京天坛祈年殿（王新喜 绘）**

  天坛是中国现存最大的古代祭祀性建筑。其建筑布局依照“天圆地方”的观念安排。天坛中轴线上的主体建筑包括南端的圜丘坛、皇穹宇和北端的祈年殿。

  The Hall of Prayer for Good Harvest in the Temple of Heaven, Beijing (Painted by Wang Xinxi)

  The Temple of Heaven is China's largest ancient building for sacrifice offering still standing today. Its architectural layout meets the concept of round heaven and square earth. The main buildings on its axle include the Circular Mound Altar and the Imperial Vault of Heaven on the south end and the Hall of Prayer for Good Harvest on the north end.

• 北京天坛祈年殿内实景图
A View inside the Hall of Prayer for Good Harvest in the Temple of Heaven, Beijing

• 福建土楼的木制门窗
Wooden Doors and Windows of the Hakka Tulou in Fujian Province

院的出入之口，它代表着尊严和等级，象征着人的“脸面”。窗比门略小，可以透光，也可以交换空气，窗框皆为木制。明清时期，门窗的装饰艺术发展迅速，蕴含着博大精深的文

structure and they give a unique style to Chinese wooden buildings.

Door and window are the building components almost solely made of wood in ancient China. The ancient Chinese took door very seriously, because it was

化意味，令人叹为观止。

用于制作家具

中国古代家具的主要用材也是木。夏商周时期朴拙的家具，春秋战国、秦汉时期浪漫的低矮家具，魏晋南北朝时期秀逸的渐高型家具，隋唐五代时期华丽的高低并存型家具，宋元时期简洁的高型家具，以及明清时期雅致的明式家具和华贵的清式家具抒写了中国传统

• 北京故宫长春宫体元殿的木窗
Wooden Windows of the Palace of Eternal Spring in the Imperial Palace, Beijing

the entrance to the entire house. Door represents dignity, social status, and the "face" of a family. Window is a bit smaller than door. It lets light and fresh air in. Window frames were all made of wood in the past. In the Ming and Qing dynasties, door and window decorative art developed rapidly and has taken with it profound cultural meanings.

For Making Furniture

Wood was also the main material for furniture making in ancient China. Traditional Chinese furniture has a long history. There had been the rough and simple furniture in the Xia, Shang and Zhou dynasties, the romantic and short furniture in the Spring and Autumn Period, the Warring States Period, and the Qin and Han dynasties, the graceful taller-and-taller furniture in the Wei , Jin, and Northern and Southern dynasties, the tall-and-short furniture in the Sui, Tang and the Five dynasties, the simplified tall furniture in the Song and Yuan dynasties, the elegant furniture in the Ming Dynasty, and the luxurious furniture in the Qing Dynasty. In the Ming and Qing dynasties, with the thriving of the imperial garden and private garden, red sandalwood and yellow rosewood were widely used for

北京故宫中和殿的木制宝座

中和殿是皇帝临时休息的场所，宝座上放有厚厚的软垫，坐着更加舒适。

The Wooden Throne in the Hall of Central Harmony of the Imperial Palace, Beijing

The Hall of Central Harmony used to serve as a temporary resting place for the emperor. The thick cushions make the throne more comfortable.

家具源远流长的历史。尤其是明清时期，皇家园林和私家园林的兴盛，使得紫檀、黄花梨等木材得以在家具制作上大放光彩，木在人们

making furniture. Wood has dominated people's lives and wooden furniture has become a first-class utensil that is of high scientific, artistic, and practical values.

的日常生活物品中占据了主导地位，木质家具成了具有科学性、艺术性、实用性的高级生活用具。

用于木雕

木雕是用木材雕刻而成的工艺品。中国木雕制作历史非常悠久，尤其是明清时期，木雕艺术大放异彩，名家辈出，留下了许多精品。在木材上，木雕一般选用质地细密坚韧、不易变形的树种，如楠木、紫檀木、樟木、柏木、红木、黄杨木等。

For Wood Carving

As a work of art, wood carving has a long history in China. It reached a great height in the Ming and Qing dynasties when famous carvers were coming forth in large numbers. They have left behind them many masterpieces. The wood varieties that are often chosen for wood carving are those with a fine texture and tough and tensile quality. They do not deform easily. These include Phoebe zhennan, red sandalwood, camphorwood, cypress wood, mahogany, boxwood, and so on.

• 清代黄杨木雕

A Boxwood Carving of the Qing Dynasty (1616-1911)

方形花几：此花几为方形，造型典雅大方。

Square flower stand: This elegant square stand is for flowerpot or vase.

长方形书画案：此书画案为长方形，外观有些像桌，但比桌精巧。

Oblong painter's table: This oblong table is finer than ordinary table, especially for art painting.

圆形坐墩：画中仕女所坐的木墩雕刻秀美，十分精致。

Rotund drum stool: The woman in the picture sits on a nicely carved wooden drum stool which is very delicate.

- **中国古代绘画中的木家具**
  Wooden Furniture in Ancient Chinese Painting

木材的其他应用

木材还有很多其他用途。在中国古代，木制品渗透到人们生活的方方面面，无论是建筑，还是桥、船、车、轿、生活用品等，皆以木为主要用材。

Other Uses of Wood

Wood has many other uses. In ancient China, it infiltrated into all aspects of people's daily life. It is the prime material for building houses and bridges and making boats, carts, sedans, and many other articles of daily use.

- 《清明上河图》（局部）[北宋 张择端]

《清明上河图》这幅传世名画描绘的是北宋汴京（今河南省开封市）清明时节市井的繁荣景象。木制品在此画中比比皆是，仅木制交通工具就有轿子、牛（马）车、人力车、太平车、平头车、木船等。此外，树木、木制建筑、木制家具也是整幅画的重要组成部分，为我们形象地展现出一个繁华的“木世界”。

Life along the Bian River at the Pure Brightness Festival (Part) [By Zhang Zeduan, Northern Song Dynasty (960-1127)]

The famous painting, Riverside Scenes at the Qingming Festival, depicts the townscape of the prosperous Bianjing City (present day Kaifeng City of Henan Province) of the Northern Song Dynasty (960-1127)at the Qingming Festival. Woodwork can be found everywhere in the painting. Wooden transport means in the painting include sedan, cattle (horse) cart, rickshaw, large cargo cart, flat-head cart, and boat. In addition, trees, wooden buildings, and wooden furniture are the main components of the entire painting, vividly presenting a prosperous world of wood to the large audience.

船：中国古代造船一般都是就地取材，根据不同部位选取不同材质。一般常用木材种类为：杉木、松木、柏木、榆木、柚木、樟木、楠木、楸木、梓木、桧木等。

Boat: Boat was built with materials taken from the neighboring areas in ancient China. Different parts of the boat were built with different kinds of wood. The commonly used wood includes those from fir, pine, cypress, elm, teak, camphorwood, Phoebe zhennan, Chinese catalpa, and juniper.

桌、椅：式样简单，材质普通，为乡村家具的代表。

Table, chair: In simple style, tables and chairs were made of ordinary materials. They were the representatives of rural furniture.

窗：样式简单、古朴，多以普通木材制成。

Window: In simple and unsophisticated styles, windows were normally made of common wood.

船舫：建设中的木制舫。舫主要仿照船的造型建在水面、岸边，用以观赏、游玩。

Fang: It was a wooden *Fang* under construction. *Fang* takes the shape of a boat and is built on water or by the bank for sightseeing and amusement.

虹桥：木桥是最早出现的桥梁形式。因其取材方便，重量轻，强度高，构造简单，常用木材有杉木、枫木、松木、柚木等。汴梁虹桥为单跨木构拱桥，由纵横梁枋相架而成，即"虹梁结构"，为中国特有的木桥形式。

Arched bridge: The earliest bridges were made of wood because wood was light, strong, and easy to get. With a simple structure, wooden bridges are normally made of the wood from fir, maple, pine, and teak. The Bianliang Arched Bridge is a single-span wooden bridge and has a crisscross structure, namely the arched beam structure, making it a unique wooden bridge masterpiece in China.

旗杆：常就地取材，以松木、桦木、杉木、杨木为主。

Flag post: Flag posts were mainly made of pine, birch, fir, and poplar wood collected from neighboring area.

轿子：像扛在肩上的木制小屋子。

Sedan: It was like a wooden cabin carried on shoulders.

房屋：宋时的木建筑结构已经非常成熟。

House: The wooden building structure in the Song Dynasty (960-1279) has already been mature.

木制车：一人拉一头驴或一匹马，车轮多采用榆木、桦木等。

Wooden cart: It was pulled by a donkey or a horse led by a carter. Its wheels were mainly made of elm and birch wood.

苏州拙政园玲珑馆内的古典家具陈设

A Classical Furniture Display in the Exquisite Hall of the Humble Administrator's Garden, Suzhou

气势恢弘的木建筑
Majestic Wooden Buildings

中国古代木建筑是世界上历史最悠久、体系最完整的建筑体系，从单体建筑到院落组合、城市规划、园林布置等在世界建筑史中都占据重要地位，体现了独一无二的“天人合一”建筑思想。经过数千年的发展，中国的木建筑形式多样，从建筑形态上可分为宫殿、园林、陵墓、民居、寺庙等。这些建筑在平面布局上多采用长方形，偶尔也采用正方形、八角形、圆形等布局方式；园林等作观赏休闲之用的建筑组合，往往采取扇形、万字形、套环形等平面组合方式，力求使整个木建筑群保持平衡对称。

The ancient wooden buildings of China form the oldest and most complete architectural system in the world. China ranks first in world architecture history in many fields, including separate buildings, courtyard compounds, city planning, and landscape gardening. They are unique in their representation of the architectural philology of unity of man and nature. Through several thousand years' development, Chinese wooden buildings have evolved into various styles, including palace, garden, mausoleum, folk house, and temple. These buildings mainly adopt an oblong plane layout and sometimes a square, octagon, and round layout. If built for sightseeing and leisure spending, the temples and gardens usually adopt plane combination modes such as sector, fylfot, and lantern ring shapes to keep the balance and symmetry of the entire wooden building.

> 宫殿

宫殿是皇帝及其后宫办公和居住的地方。宫殿一般规模宏大，布局规范严谨，以突显皇权的尊贵。中国留存下来的宫殿大都是木制的，由于战争及朝代更迭等原因，很多宫殿被毁，如今完整保存下来的帝王宫殿群有北京故宫和沈阳故宫。

北京故宫

北京故宫是中国明清两代的皇宫，始建于1406年，规模宏大，气势磅礴，布局严谨，是世界上现存皇宫中历史最悠久、建筑面积最大、保存最完整的一座，也是世界上现存最大、最完整的木质结构古建筑群。

故宫所用的建筑材料主要是木材和砖石，木材以金丝楠木为主，且工艺精湛、做工精细，体现出森

> Palace

Palace is the residence and office of emperor and his families. Normally in a grandiose scale and a standardized layout, the palace represents the supreme status of the imperial power. The remaining palaces in China were almost all made of wood. Many palaces have been destroyed during the war or the change of dynasties. Now, the imperial palace complexes that have been well-preserved in China include the Imperial Palace in Beijing (the Forbidden City) and the Imperial Palace in Shenyang.

The Imperial Palace in Beijing (the Forbidden City)

The Imperial Palace in Beijing served two dynasties in Chinese history: Ming Dynasty (1368-1644) and Qing Dynasty

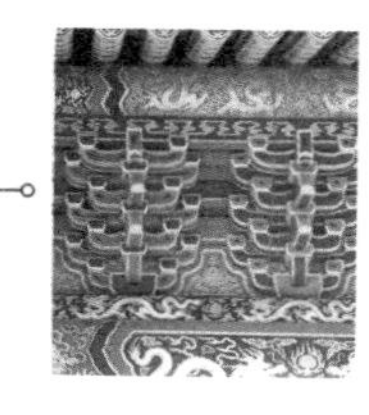

柏木、楠木、樟木制作的斗拱（斗拱：中国建筑特有的一种承重结构，设在立柱和横梁交接处。从柱顶上一层层探出呈弓形的承重结构叫“拱”，拱与拱之间垫的方形木块叫斗）

Bucket arches made of cypress, Phoebe zhennan, and camphorwood (Note: Bucket arch is a unique weight-bearing structure in Chinese architecture. It is set at the joint between the supporting column and crossbeam. The bow-shaped weight-bearing structure extending from the top of the column is called arch and the square block between the arches is called bucket.)

楠木制作的门窗

Doors and windows made of Phoebe zhennan wood

楠木制作的梁、柱

Beam and column made of Phoebe zhennan wood

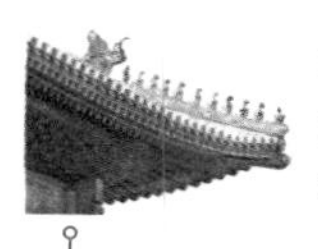

松木制作的屋檐

Eaves made of pine wood

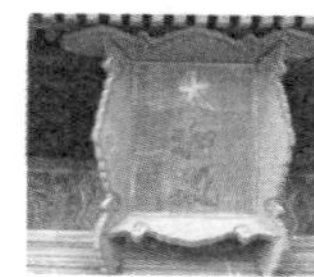

太和殿木制匾额（匾额：中国古建筑物的标识，一般挂在门上方、屋檐下）

The wooden inscribed board bearing the name of the Hall of Supreme Harmony (Note: The inscribed board was the sign of ancient Chinese buildings. It is normally hung above the door and below the eaves.)

北京故宫太和殿

太和殿空间高敞开阔，是中国现存最大的木结构大殿，高26.92米，建筑面积达2377平方米。太和殿共有72根巨大的柱子支撑着沉重的屋顶，中央有6根金光闪闪的蟠龙柱环绕金銮宝座四周，屋顶上有一个巨大的金漆蟠龙。

The Hall of Supreme Harmony in the Imperial Palace (the Forbidden City), Beijing

The Hall of Supreme Harmony has a high, spacious, and open space. Some 26.92 meters tall and with a floorage of 2,377 square meters, it is the largest standing wooden hall in China. The Hall of Supreme Harmony has 72 massive columns to support its heavy roof. In the center of the hall are six glittering columns with curled-up dragons and they circle around the golden throne. Clinging to the roof is a large curled-up dragon coated with gold lacquer.

严的等级。故宫整体通过建筑的空间变化和装饰细节展现出宏伟壮丽的构思和皇权至尊观念。

楠木

在中国古典建筑中，楠木一直被视为最理想、最珍贵、最高级的建筑用材，是皇家宫殿及重要建筑的用材。楠木纹理美观，结构致密，质地温润柔和，切面有光泽，还散发出阵阵幽香，具有很强的防潮防腐蚀性。北京故宫太和殿、长陵以及北海公园内的大慈真如宝殿、快雪堂等宫廷建筑都有楠木构件，也有楠木宝座、箱柜等家具。

Phoebe zhennan Wood

Phoebe zhennan has always been regarded as the ideal, most precious, and first-class material for building imperial palace and other important structures in Chinese classical architecture. It has beautiful wood grains, a compact structure, and a mild and gentle texture. Its section looks glossy and sometimes sends forth fragrance, indicating a superb resistance against moisture and corrosion. Phoebe zhennan components can be found in many imperial buildings in Beijing, including the Hall of Supreme Harmony in the Imperial Palace (the Forbidden City), Changling Mausoleum (one of the famous Tombs of the Ming Dynasty), and the Hall of Great Mercy and Eternal Truth and the Hall of Quick Snow in Beihai Park. Among them are thrones, cases, and other kinds of furniture.

(1616-1911). Built in 1406, it has a large scale, an imposing manner, and a well-knit layout. It is the oldest, biggest, and best-preserved imperial palace in the world. It is also the largest and the most complete ancient wooden building group in the world.

The Imperial Palace was built mainly with wood, bricks, and stone slabs. The precious gold wire Phoebe zhennan was used. The palace shows great skill, refined craftsmanship, and strict hierarchy. Through change of architectural space and display of decorative details, the palace fully demonstrates the majesty and supremacy of the imperial power.

• 楠木的心材
Phoebe Zhennan Heartwood

- 北京故宫太和殿内部的陈设

清代，随着红松数量的减少，大量使用包镶、拼接等技术，太和殿内高13米的“金龙柱”就是这样拼合而成的。

Interior layout of the Hall of Supreme Harmony in the Imperial Palace (the Forbidden City), Beijing

In the Qing Dynasty (1616-1911), with the decrease of Korean pine, builders widely adopted techniques such as wrapping, inlaying, and splicing. The 13-meter golden dragon column in the Hall of Supreme Harmony was put together in this way.

松木

松木的心材和边材区别明显，心材呈红褐色，边材呈浅红褐色，生长轮非常明显，纹理直或斜，具有松香味。松木家具或建筑在视觉上以“色”夺人。中国古典松木家具以突出其自然性为主，造型古朴大方，线条简洁明了，风格朴实含蓄、严密雅致。

Pine Wood

The heartwood and sapwood of pine have obvious differences. The heartwood is reddish brown while the sapwood is lighter. The growth ring of pine tree is apparent. There are straight or skew grains and rosin fragrance. The furniture or buildings made of pine wood have eye-catching colors. Chinese classical pine wood furniture is known for its natural feature, simple and unsophisticated profile, clear-cut lines, and reserved, compact, and elegant style.

- 红松木的皮
  Bark of Red Pine

- **北京故宫储秀宫的雕花木罩**
  雕花木罩是用来分隔室内空间的，一般安装在前、后檐金柱之间，同时具有很好的装饰作用。
  The Carved Wooden Casing in the Palace of Gathering Elegance of the Imperial Palace (the Forbidden City), Beijing
  The carved wooden casing is for partitioning the indoor space. A very good decorative component, it is normally installed between the golden columns supporting the front and rear eaves.

- 北京故宫乾清门外的木制匾额

故宫中的匾额皆木制，分横、竖两类。此匾为竖匾，即竖长方形，蓝底金字，四边叠起雕龙花板，其上书写宫门、殿名，悬挂于宫门、殿堂的檐下。

The Wooden Inscribed Board outside the Gate of Heavenly Purity in the Imperial Palace(the Forbidden City) ,Beijing

All the inscribed boards in the Imperial Palace , horizontal or vertical, were made of wood. This is a vertical one bearing the Chinese name of the Gate of Heavenly Purity. Its blue background is encircled by the boards with dragon carving, on which the gold characters are written. Other boards in the Imperial Palace also bear the names of the corresponding gates and palaces and are hung under the eaves.

- 北京故宫太极殿院内的木影壁

影壁是集实用性和装饰性为一体的一种建筑装饰。这座木影壁框为木制，饰朱红油漆，壁心是金色的五蝠捧团寿字，四周为象征“洪福齐天”的群蝠流云纹，四角亦是金色的蝠纹，颜色和图案都非常华美，寓意吉祥、长寿。

The Wooden Screen Wall in the Courtyard of the Hall of Supreme Principle in the Imperial Palace (the Forbidden City), Beijing

Screen wall is a practical and decorative structure. This wooden screen wall has a wooden frame coated with red paint. At its center are five golden bats (bat in Chinese sounds like good fortune) forming a character representing longevity. On the edge of the screen are the bats and flowing cloud pattern representing great luck and fortune. There are also golden bat patterns on four corners. They are extravagantly beautiful and indicate luck and longevity.

沈阳故宫

沈阳故宫始建于1625年，是清朝前期建造的皇宫，是中国现存仅次于北京故宫的最完整的皇家建筑群。沈阳故宫因其独特的地域特色和浓郁的满族风情而迥异于北京故宫。其中，金龙蟠柱的大政殿、八字形排列的十王亭、“筒子房”格局的清宁宫、古朴典雅的文朔阁以及凤凰楼等高台建筑，在中国宫殿建筑史上都是绝无仅有的。

The Shenyang Imperial Palace

The Shenyang Imperial Palace was built in 1625. It is the second most complete imperial palace in China, second only to that in Beijing. Shenyang palace is totally different from Beijing palace in its unique regional characteristics and rich Manchu style. Many of its high-ground buildings are rare masterpieces in Chinese palace architectural history, including the Hall of Great Government with golden dragon columns, the Pavilion of Ten Princes in a splayed layout, the Palace of Pure Tranquility in a tube-shaped pattern, the simple and unsophisticated Wenshuo Pavilion, and the Phoenix Building.

沈阳故宫的凤凰楼

凤凰楼是沈阳故宫后宫——清宁宫的大门楼，共三层。凤凰楼的周围有廊，内部雕梁画栋，当年是商议军政大事和举行宴会的场所。凤凰楼上彩绘有形态各异的龙的纹样，富丽堂皇，颇具皇家气派。

The Phoenix Tower of the Shenyang Imperial Palace

The Phoenix Tower is a three-storey gateway arch of the Palace of Pure Tranquility, the harem of the Shenyang Imperial Palace. It is surrounded by corridors and inside it, there are carved beams and painted rafters. It served as the venue for the emperor to discuss state affairs with his officials and to give banquets. The outer walls of the building are covered with dragon patterns in all forms and shapes. It fully demonstrates the extravagance and grandeur of the royal family.

• 沈阳故宫俯视图（图片提供：图虫创意）
A Vertical View of the Shenyang Imperial Palace

中国木建筑的梁柱结构

中国古人建造的房屋主要有两种梁柱结构，即抬梁式和穿斗式。抬梁式是指在前后大柱之间架设大梁，用来承受屋顶的重量。大梁上重叠数层小梁，逐层缩短，梁间用矮柱托垫，构成两面斜坡的屋架。抬梁式房屋主要流行于中国北方地区。穿斗式是在各大柱中水平地穿过若干根木枋，使之联成一体。穿斗式房屋主要流行于中国的华东、华南地区。

Beam Column Structure of Chinese Wooden Building

The ancient Chinese built houses mainly in two kinds of beam-column structures: post and lintel construction & column and tie construction. In post and lintel construction, a crossbeam is set between the front and rear posts to carry the weight of the roof. The crossbeam carries several layers of minor beams and their length dwindles layer by layer. Short pillars are inserted between the beams to form a roof truss with two slopes. The post and lintel construction is popular mainly in North China. In column and tie construction, several tie beams are inserted horizontally through the major columns to bind them up. The column and tie construction is popular mainly in East China and South China.

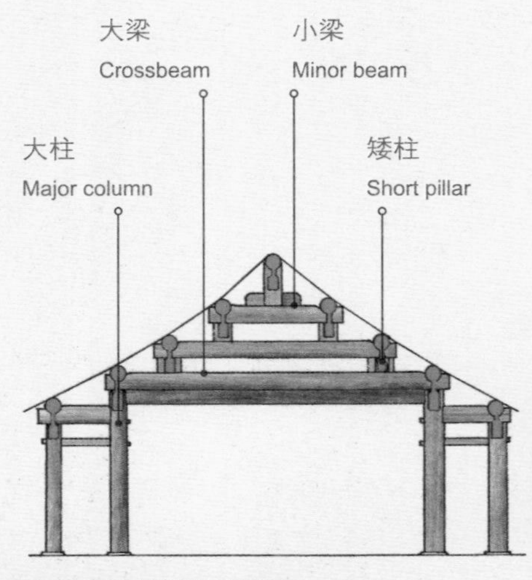

• 抬梁式房屋结构示意图
A Schematic Diagram of the Post and Lintel construction

• 穿斗式房屋结构示意图
A Schematic Diagram of the Column and Tie Construction

> 园林

中国的园林历史悠久、风格独特，主要有皇家园林、私家园林、官署花园、寺院园林、公共园林等类型。园林中的主要建筑类型包括殿、堂、亭、楼、阁、廊、厅、轩、馆、榭、舫等，皆以木石为材料建成。与其他建筑相比，园林建筑更加注重观赏性及与园林景观的和谐、呼应。

颐和园

颐和园是中国保存最完整的皇家园林，占地面积2.9平方千米，汇集了中国传统造园艺术的精华。颐和园气势恢弘，富丽堂皇，又充满了自然之趣，在中外园林艺术史上地位显著。园内建筑以佛香阁为

> Garden

Chinese garden has a long history and a unique style. There are imperial garden, private garden, public garden, and garden in official mansion and temple. Inside the garden are mainly the structures built with wood and stone, including palace, hall, pavilion, storied building, corridor, veranda, shed, *Fang*, and so on. Different from other buildings, garden structures are built to add beauty, harmony, and elegance to the garden.

The Summer Palace

The Summer Palace is the best-preserved imperial garden in China. Occupying a land area of 2.9 square kilometers, it houses the essence of the traditional landscape gardening of China. The Summer Palace is majestic, luxurious,

中心，大小建筑百余座、院落20多处。颐和园以昆明湖形成的水面为特色，以楼、阁、塔、台等木建筑为呼应和点缀。

and full of fun the nature can offer. It takes an important position in the history of landscape art at home and abroad. Over 100 buildings and 20 courtyards in all sizes are scattered in the palace and they all center on the Pavilion of Buddhist Incense. A major feature of the Summer Palace is the Kunming Lake, which is surrounded and decorated by many wooden structures such as storied buildings, towers, pagodas, and terraces.

- **颐和园中的木制长廊**

廊是园林中各个建筑之间的交通要道，同时又是风景导游线，可游可憩，还是很好的风景。廊的造型丰富多样，常常弯曲狭长，依势而曲，可蜿蜒山坡，凌空水上。

The Wooden Long Corridor in the Summer Palace

Corridors are passageways linking all buildings in a garden. They are also the ideal sightseeing route with resting facilities. Corridors have rich and diversified shapes. Long and narrow, they often wind their way through the terrain, some up the hills and some across the water.

- **颐和园宜芸馆内的陈设**

馆具有游宴听曲、起居会客的作用，在园中位置一般不显著，规模有大有小，布置较灵活，或面向庭院，或临水倚楼。

Interior Layout of YiYun Hall in the Summer Palace

Halls are for giving banquets and opera performances, spending leisure hours, and meeting guests. They do not usually take the conspicuous position in a garden. Their sizes, and layouts are flexible. Some of them face the courtyard while others face the water or stand by storied buildings.

亭

亭是园林中用得最多的游赏建筑，可供人们休息、观景，遮阳避雨，同时还是园林风景的重要点缀。亭以造型小巧秀丽、玲珑多姿为特色，布局十分灵活，可建于山上、林中、路旁、水上。亭子的体量虽然不大，但造型变化却非常多样。亭子的形状一般有正方、长方、三角、六角、十字、圆形、梅花、扇面等。按位置分，亭又有山亭、半山亭、桥亭、水亭，靠墙的半亭、路中的路亭等。

Pavilion

The pavilion is the most-common building in the garden. It serves as a place for rest and sightseeing and as a shelter from the sun and the rain. It is also an important decoration for the garden. Pavilions are often small and exquisite and can be built on the mountain, in the woods, by the road, and over the water. Though small, they vary in shapes. They normally take the shape of a square, an oblong, a triangle, a hexagon, a cross, a round, a plum blossom, or a sector. Classified by their location, there are mountain pavilion, mountainside pavilion, bridge pavilion, and water pavilion. The one beside a wall is a half pavilion and the one in the road is a road pavilion.

• 颐和园"意迟云在"敞厅内部的木质结构
The Wooden Structure inside the Open Hall of Late Intention and Staying Cloud in the Summer Palace

• 颐和园含新亭
The Pavilion of Novelty in the Summer Palace

• **颐和园中的佛香阁**

阁是园林中常用的建筑形式之一，一般依山临水而建，临水而建的阁又称“水阁”。阁与楼常连在一起，但跟楼相比，其造型更为轻巧、通透，四面开窗；平面常呈四方形或对称多边形，比如六边形、八边形等。佛香阁是一座八边形的木构佛殿，坐北朝南，高41米，坐落在方形花岗石台基上。

The Tower of Buddhist Incense in the Summer Palace

A tower(*Ge*) is a commonly-used architectural form in gardens. It is normally built beside a mountain and facing the water. Those facing the water are called water towers. Tower is often connected with storied buildings and in comparison, it is lighter and opener, and has windows at all four sides. Normally, pavilions take the shape of a square or a symmetrical polygon such as hexagon and octagon. The Tower of Buddhist Incense is a wooden octagon hall. Situated on the north and facing the south, the 41-meter-tall tower is built on a square granite platform.

北海公园

北京北海公园是中国现存最古老、最完整、最具综合性和代表性的皇家园林之一。北海原本是辽、金、元代的离宫，后来成为明清帝王的御苑。全园以北海为中心，面积约71公顷，水面开阔，湖光塔影，亭台别致，游廊曲折，犹如仙境一般。园内有画舫斋、镜清斋、天王殿、五龙亭等著名建筑。

Beihai Park

Beijing Beihai Park is one of the oldest, most complete, and most comprehensive and representative imperial gardens in China. It was originally the palace for the emperors of the Liao, Jin, and Yuan dynasties during their tours out of the capital. Later, it became the imperial garden in the Ming and Qing dynasties. The Beihai Lake, some 71 hectares in area, is the center of the park. Surrounded by exquisite pavilions, terraces, and corridors, the lake makes the park a fairyland in the mundane world. Other famous buildings in the park include the Painted Boat Hall, Mirror Study, Hall of Heavenly Kings, and Five-dragon Pavilion.

- **北海公园的栏杆**

栏杆是中国古建筑的重要木构件，在楼层平台、长廊、亭榭临边之处都设有栏杆。除了安全防护功能外，栏杆还发挥着装饰作用。

Carved Balusters in Beihai Park

Baluster is an important wooden component in ancient Chinese buildings. It is normally set on the edge of building platforms, long corridors, and pavilions. In addition to its protective function, it also plays a decorative role.

中国四大名园

1961年，国务院颁布的第一批全国重点文物保护单位名单中，属于园林方面的有四处，这四处园林因此而被称为“中国四大名园”。它们分别是江苏省苏州市的拙政园、北京市的颐和园、河北省承德市的避暑山庄和江苏省苏州市的留园。

Top Four Gardens of China

In 1961, the State Council of China issued the name list of the first group of the key cultural relics under national protection. Four gardens were among the relics and thus they have been hailed as Top Four Gardens of China. They are the Humble Administrator's Garden in Suzhou City of Jiangsu Province, the Summer Palace in Beijing City, the Mountain Resort in Chengde City of Hebei Province, and the Lingering Garden in Suzhou City of Jiangsu Province.

龙泽亭中的蟠龙藻井

The dragon caisson ceiling in the Pavilion of Dragon Benevolence

- **北海公园里的龙泽亭**

龙泽亭上圆下方，旧时，只有皇帝才可以在这里垂钓。

The Pavilion of Dragon Benevolence in Beihai Park

The Pavilion of Dragon Benevolence has a round dome and a square foundation. It was a fishing ground exclusively for the emperor himself.

• 北海公园里的爬山廊
The Hill Corridor in Beihai Park

• 北海公园延楼长廊近景
A Close-up View of the Yanlou Corridor in Beihai Park

门上的云纹：门窗上雕有精美的浮云图案。

Cloud pattern on a door: This door has exquisite cloud pattern on it.

楠木的木纹：门梁长期经受风吹雨淋，木质依然完好，没有腐蚀的痕迹。

Wood grain of Phoebe zhennan: Despite a long-time exposure to wind and rain, this door beam is still in a good condition and shows no sign of corrosion.

楠木巨柱：整座大殿由20根高达10米、直径半米的楠木巨柱支撑。

Huge Phoebe zhennan columns: The entire hall is supported by 20 huge Phoebe zhennan columns. Each of them is 10 meters in height and half a meter in diameter.

木窗：门窗做工华丽，装饰有繁复考究的菱形窗格。

Wooden windows: These wooden windows are beautifully made and decorated with complex and exquisite rhombus lattices.

宝殿的匾额

The inscribed board of a hall

- **北海大慈真如宝殿**

大慈真如宝殿，又称“楠木殿”，是国内极为罕见的金丝楠木大殿。整座建筑全部采用名贵的金丝楠木建造且不施彩绘，十分雍容古朴。

The Hall of Great Mercy and Eternal Truth in Beihai Park

Also known as the Phoebe Zhennan Hall, the Hall of Great Mercy and Eternal Truth is a rare building of its kind in China. It was made of the precious gold wire Phoebe zhennan and no painting was ever done. This gives it a graceful and unsophisticated look.

• 北海公园的静心斋

斋是园林中幽深僻静处的学馆书屋，一般空间比较封闭，往往藏而不露，形式不拘，借外景，自然幽雅，深得山林之趣。

Heart-ease Study in Beihai Park

A garden study is often situated at a secluded place. It has a closed space and diversified forms. It fits in well with the surrounding natural environment.

承德避暑山庄

河北的承德避暑山庄是中国现存最大的古典皇家园林，有殿、堂、楼、馆、亭、榭、阁、轩、斋、寺等建筑一百余处，是中国三大古建筑群之一。承德避暑山庄的

The Mountain Resort in Chengde

The Mountain Resort in Chengde, Hebei Province is the largest existing classical imperial garden in China. It houses over 100 palaces, halls, storied buildings, pavilions, verandas, studies, and temples, making it one of the top three ancient

• 承德避暑山庄的烟雨楼

园林中的楼一般为二层及以上的观赏建筑，可登高远眺，游憩观景。楼主要由木材搭建，多设于园林周边，也常依山傍水而建，成为风景的主体。

Yanyu Lou (House of Mists and Rains) in the Mountain Resort in Chengde

The buildings in a garden are normally two-storey or higher structures for sightseeing. They are mainly built with wood on the edge of the garden. Sometimes, they stand beside hill and water and serve as the main part of the scenery.

• 承德避暑山庄的含澄景亭

Hanchengjing Pavilion in the Mountain Resort in Chengde

木建筑均采用原木本色，淡雅庄重，简朴自然，展示了中国古代木结构建筑的高超技艺，也实现了木结构与砖石结构的完美结合。

拙政园

在中国古代园林中，除皇家园林外，还有私家园林，即王公贵族以及士大夫、富商等修建的私人园林。

在繁花似锦的私家园林中，以苏州园林最为著名。苏州园林大多面积不大，但以意境见长，拥有各式典雅别致的木建筑。在众多的私家园林中，拙政园被公认居于首

building groups in China. Wooden buildings in the Mountain Resort in Chengde all bear the true color of the log. They have elegant, solemn, simple, and appropriate appearances, fully demonstrating the outstanding skills of the builders of ancient Chinese wooden construction. They also represent the perfect combination of wooden structure and masonry structure.

The Humble Administrator's Garden

Among ancient Chinese gardens, there are private ones in addition to imperial ones. They were built by princes, nobles, officials, and rich businessmen.

Suzhou boasts the most famous private gardens, hence the well-known term Suzhou Garden. Though small, these gardens excel in artistic conception. They also house elegant wooden buildings. Of the numerous private gardens, the Humble Administrator's Garden is recognized as

拙政园浮翠阁

浮翠阁是一座八角形双层木建筑，周围开窗，掩映在绿树之中，看起来像是浮动在绿荫之中，因此被称为“浮翠阁”。

The Floating Green Pavilion in the Humble Administrator's Garden

The Floating Green Pavilion is an octagon two-story wooden building with windows at all four sides. It stands in green trees and looks like a pavilion floating in the green.

位。该园以水为中心，池水面积约占总面积的五分之一，各种亭、台、轩、榭等木建筑多临水而建，非常精巧秀丽。

the best. It has a pond, almost one fifth of its entire size, at the center. Around the pond are various elegant and exquisite wooden structures such as pavilions, terraces, and verandas.

• **拙政园见山楼**

见山楼是拙政园中部水池西北角的一座水上楼阁，三面环水，两侧傍山，门窗梁架均为木制，从西部可通过平坦的廊桥进入底层，而上楼则要经过爬山廊或假山石级。

The Mountain in View Tower in the Humble Administrator's Garden

The Mountain in View Tower is a waterborne pavilion at the northwest corner of the central pond in the Humble Administrator's Garden. It faces water on three sides and is accompanied by hills on two sides. Its doors, windows, beams, and skeletons are all made of wood. A visitor can enter its bottom storey from the west through an even corridor bridge. Its upper storey is accessible through the hill corridor or rockery steps.

舫

舫是修建在中国园林水池畔的船形建筑，舫的前半部多深入水中，给人以置身舟中的感觉，舫首一般设有桥供人进入舫中。舫多由前舫、中舫、尾舫三部分组成，前舫较高，舫首开放，可供人赏景；中舫较低，多设有矮墙或窗，是供人休息或设宴的地方；尾舫最高，多为两层且四面开窗以供人远眺。中国江南私家园林中的舫多仿湖中画舫修建，造型灵活多变。

Fang

Fang is a boat-shaped structure built by the pond in Chinese gardens. Its front part normally stands in the water, giving the visitors a sense of staying in a real boat. A *Fang* is composed of three parts: bow, middle cabin, and stern. At the bow, there is usually a bridge as the access to the boathouse. The front part of the *Fang* is relatively high and it is an open space for sightseeing. The middle part is lower. With parapets or windows, it is for rest or giving banquets. The rear part is the highest with a two-storey structure for the visitors to enjoy the views in the distance. The *Fang* of the private gardens in the regions south of the Yangtze River are mostly the copies of the painted boat in the lake. They show flexible and diversified shapes and profiles.

- 拙政园卅六鸳鸯馆

卅六鸳鸯馆曾是拙政园主人会客、休憩、宴请的地方。此馆几乎全由木制，顶部采用弧形设计，便于反射声音，使馆内的音响效果奇好。

The 36 Pairs of Mandarin Ducks' Hall in the Humble Administrator's Garden

The 36 Pairs of Mandarin Ducks' Hall served as a venue for meeting guests, resting, and giving banquets. The entire hall is almost made of wood only. Its curved ceiling is good for sound reflection, hence an extremely good acoustic effect in the hall.

- 拙政园香洲舫

香洲舫的船头是台，前舱是亭，中舱为榭，船尾是阁，阁上起楼。两层楼舱均为木制，雕梁画栋，典雅秀美。

Fang of Fragrance Islet in the Humble Administrator's Garden

The bow of the *Fang* of Fragrance Islet is a platform. Its fore-cabin, middle cabin, and stern are three pavilions in different styles. On the pavilions are storied buildings made of wood. With carved beams and painted rafters, the buildings are very elegant and graceful.

留园

留园是苏州诸园中木建筑数量最多的园林，有“吴中第一名园”的美誉。园林内布满了厅堂、走廊、粉墙、洞门、假山、水池等建筑，还有数不清的花草树木。留园的木建筑虽然多，但空间布局十分合理，建筑与假山、水池、花木等

The Lingering Garden

The Lingering Garden boasts the largest number of wooden buildings among all Suzhou gardens. In fact, it is dubbed the No.1 Garden in the center of Wu Area. It houses numerous halls, corridors, whitewashed walls, moon gates, rockeries, and ponds, and countless flowers and grasses. Though in large numbers, the wooden buildings in the Lingering Garden are in a reasonable spatial layout. They form a seamless whole with the surrounding rockeries, ponds, and plants, turning the garden into a building group with rich and well-arranged structures. It fully represents the refined skills and excellent

• **留园一景——小蓬莱**

留园小蓬莱是一组木制凉亭，通过曲桥相连，上面架以亭式紫藤棚架。

The Small Penglai Pavilion, a Scenic Spot in the Lingering Garden

The Small Penglai Pavilion in the Lingering Garden is a group of wooden arbors connected by the winding bridge and covered with pavilion-style vine shelves.

有机地融合在一起，形成一组组层次丰富、错落有致的建筑群，充分体现了古代造园家的高超技艺、卓越智慧以及江南园林建筑的艺术风格和特色，堪称中国私家园林建筑中的精品。

wisdom of ancient landscape architects, and the artistic style and characteristics of the garden buildings in the regions south of the Yangtze River. It is indeed a masterpiece among the private gardens of China.

留园冠云亭

The Guanyun Pavilion in the Lingering Garden

留园五峰仙馆内部的陈设

五峰仙馆俗称“楠木厅”，梁柱全部采用楠木，中间全部采用红木银杏纱隔屏风，厅内装饰精丽，陈设雅洁大方，是江南厅堂的典型代表。

Interior Layout of the Celestial Hall of Five Peaks in the Lingering Garden

The Celestial Hall of Five Peaks is also called the Phoebe Zhennan Hall due to the fact that all of its beams and columns are made of Phoebe zhennan. In the middle of the hall is a screen made of mahogany and gingko wood. The hall is elegantly decorated and gracefully arranged, a typical representative of halls in the regions south of the Yangtze River.

> 民居

民居就是普通百姓的居住之所，中国的民居所用建筑材料大多为土、砖石和木材。由于中国疆域辽阔，又是一个多民族组成的大家庭，因不同的地理条件、气候条件

• 江苏周庄古镇民居二层的木窗
Wooden Windows on the Second Floor of a Folk House in the Old Town of Zhouzhuang in Jiangsu Province

> Folk House

Folk house is the residence of ordinary people. Chinese folk houses are normally built with earth, brick, stone slabs, and wood. China is a large country with many ethnic groups. Different geographical and weather conditions, different life styles, different economic and cultural influences give rise to different house patterns and styles. Typical traditional Chinese folk houses include the folk house in the regions south of the Yangtze River, folk house in Northwest China, Beijing folk house, folk house in South China, and folk house of ethnic groups.

Folk House in the Regions South of the Yangtze River

In the regions south of the Yangtze River, folk houses are mostly built beside water

以及不同的生活方式，再加上经济、文化等各个方面的影响，各地民居房屋样式以及风格有所不同。中国有特色的传统民居建筑包括江南民居、西北民居、北京民居、华南民居以及少数民族民居等。

because water transport there is highly developed. Southern area of China is also subject to a hot, humid, and rainy climate. To avoid humidity, local folk houses normally have high walls and

江南民居

江南民居大多临水而建，因为古代江南地区水运相当发达，南北货运主要依靠水运。此外，中国南方炎热、潮湿、多雨，为了防潮避湿气，江南民居的墙一般较高大，开间也大，设前后门，便于通风。为了隔绝地上的湿气，一般为两层建筑，底层多为砖墙，上层为木结构，内部结构多为穿斗式木构架，即用枋把柱子串联起来，形成一层层房架。厅堂内部多用木制的罩或屏门等分隔。

江苏周庄古镇的民居

周庄古镇依河成街，桥街相连，河埠廊坊，临河水阁，处处显得古朴幽静，是典型的江南古镇。全镇有近百座古典宅院和14座各具特色的古桥，宅院下层为砖墙，上层为木制，非常有特色。

A Folk House in the Old Town of Zhouzhuang in Jiangsu Province

The old town of Zhouzhuang also has a riverside street, which is linked with bridges. Along the river are docks, corridors, arches, and water pavilions. It is a typical quiet and unsophisticated old town to the south of the Yangtze River. The whole town has nearly one hundred classical courtyards and 14 ancient bridges with different characteristics. Here, the folk houses have brick-wall ground floor and wooden-structure upper floor.

• **江南西塘古镇的民居**

西塘古镇位于江浙沪交界处，临河的街道都建有廊棚，总长近千米。西塘古镇下层的砖墙较高，上层的木制建筑较低矮。

Folk Houses in the Old Town of Xitang to the South of the Yangtze River

The old town of Xitang is located at the juncture among Jiangsu Province, Zhejiang Province, and Shanghai. The riverside streets in the town are lined with corridor sheds extending as long as one thousand meters. The ground-floor brick walls in Xitang are relatively tall and the upper-floor wooden structures are relatively short.

徽州民居

徽州民居主要分布在安徽省各地，主要特点是墙体高大，能把屋顶都遮挡起来，起到防火的作用。门楼用石雕、木雕和砖雕进行装饰，装饰纹样富有生活气息。宅院大多依地势而建，分三合院、四

large rooms with front and back doors for better ventilation. In addition, to get rid of the dampness from the ground, the houses are usually two-story buildings, their ground floor with brick walls and upper floor in wooden structure. They have a column and tie construction interior. Tie beams are used to link the columns and form layer upon layer of house stands. The interior of the hall is partitioned with wooden casing or screen.

Huizhou Folk House

Huizhou folk houses are mainly distributed in Anhui Province. Their walls are so tall that they can cover the roof as a fire-control measure. The gateway arch is decorated with stone carving, wood carving, and brick carving. The decorative patterns are full of the taste of real life. The courtyards are mostly built by following the terrain. There are three-sided courtyards and four-sided courtyards. The gate of the Huizhou folk house is on the axle line of the entire house. Standing in the center of the house is the main hall. The backyard has a two-story building. The four-sided courtyard is called the patio, which is for lighting and drainage.

合院。徽州民居的大门位于中轴线上，正中为大厅，后面院内有二层楼房，四合房围成的小院称为“天井”，天井有采光和排水之用。

江西婺源江湾萧江宗祠
Xiao and Jiang's Ancestral Hall in Jiangwan, Wuyuan, Jiangxi Province

婺源古村落

婺源古村落位于江西省东北部，这里自然环境奇佳，峰峦、幽谷、溪涧、林木、奇峰、异石、古树、驿道、亭台、廊桥、溶洞奇多，是我国现今保存数量最多、最完好的古建筑群之一。此地有明清时期的古祠堂113座、古府第28栋、古民宅36幢、古桥187座，几乎遍布全县各乡村。婺源民居以石雕、木雕、砖雕“三雕”著称，用材考究，做工精美，风格独特，有着深厚的文化底蕴。其中，被称为“木雕宝库”的汪口俞氏宗祠，占地665平方米，其梁、柱、窗上的浅雕、深雕、浮雕、透雕形成各种图案达100多组，刀工细腻，工艺精湛。

Wuyuan Old Villages

Wuyuan old villages are located in the northeast of Jiangxi Province. Enjoying an ideal natural environment, the villages are scattered in all the townships of the county. They are surrounded by mountain ranges, deep valleys, springs and waterfalls, woods, towering peaks, strange rocks, ancient trees, courier routes, pavilions, corridor bridges, and karst caves. They house one of the largest best-preserved old building groups in China, including 113 ancestral halls, 28 mansions, 36 folk houses, and 187 bridges built in the Ming and Qing dynasties. Wuyang folk houses are famous for three carvings, namely stone carving, wood carving, and brick carving. Built with well-chosen materials, they show refined workmanship, unique style, and profound cultural background. Among them, Yu's ancestral hall in Wangkou is the most famous. Dubbed the No.1 ancestral hall in the regions south of the Yangtze River, it occupies an area of 665 square meters and boasts over 100 groups of carving patterns on its beams, columns, and windows. These carvings include low relief, high relief, and fretworks, all showing refined and excellent workmanship.

徽州民居

徽州的明清古民居总计有7000栋，且形式多样，五花八门，约有十五种之多，如古城、古村镇、祠堂、寺庙、书院、园林、戏台、牌坊、桥梁、塔、亭、堤坝、井等。这些建筑多为木制，且每处建筑上都有雕刻精美的纹饰。

A Huizhou Folk House

In Huizhou, there are totally 7,000 old folk buildings. Built in the Ming and Qing dynasties, they can be divided into about 15 types, including old town, old village, ancestral hall, temple, seminary, garden, opera stage, archway, bridge, pagoda, pavilion, dike, and well. They are mostly wooden buildings with exquisite carving.

徽州民居门楼斗拱上的木雕

Wood Carving on a Bucket Arch of a Huizhou Folk House

晋中民居

晋中即山西省中部，这里有北方传统民居合院形制的典型代表，最出名的是晋商们修建的豪宅大院。这些民居大多修建于清代，建筑规模较大，设计精巧，具有独特的建筑造型和空间布局。晋中民居一般为砖木结构，且砖墙多为清一色的青砖，墙体厚实。内部构架多为木制。晋中一些大规模的民居建筑，如著名的乔家大院、王家大院等是晋中民居的代表。

• 乔家大院木雕
Wood Carving in the Qiao's Grand Courtyard

Folk House in the Center of Shanxi Province

The central area of Shanxi Province has many typical representatives of traditional courtyard-style folk houses commonly-seen in North China. Among the most famous are the luxurious courtyard houses built by Shanxi businessmen. These folk houses were mostly built in the Qing Dynasty (1616-1911). They have large scale, exquisite design, and unique architectural shapes and spatial layout. The folk houses in the area are normally in post and panel structure. Their thick and sturdy walls were built with black bricks only. The houses have wooden interior structures. The famous large-scale folk houses in the center of Shanxi Province include the Qiao's Grand Courtyard and Wang's Grand Courtyard, both the representative of local folk houses.

• 王家大院房屋顶部的木结构
The Wooden Roof of a Room in the Wang's Grand Courtyard

- **山西乔家大院院落**

乔家大院是清代著名商人乔致庸的宅第，始建于乾隆年间，后增修和扩建。乔家大院被认为是中国清代北方民居风格的典型代表，共有6个大院，20个小院，建筑布局规整，富丽堂皇，全院的亭台楼阁都有精美的木雕装饰，是中国古代民居建筑中不可多得的精品。

The courtyard of the Qiao's Grand Courtyard in Shanxi Province

The Qiao's Grand Courtyard was the mansion of the famous businessman Qiao Zhiyong in the Qing Dynasty. It was originally built during the reign of Qianlong and was repaired and expanded in the following years. The Qiao's Grand Courtyard is regarded as a typical representative of northern folk houses of China built in the Qing Dynasty (1616-1911). With 6 large courtyards and 20 small ones, the compound is in a well-knit layout and shows majestic splendor. All of its pavilions, terraces, and buildings bear exquisite wood carving decorations, making it a rare masterpiece among old folk houses of China.

- **山西王家大院的院门**

王家大院是王氏家族从清康熙年间开始修建，后陆续增建而成的宅第，规模宏大。其建筑分“五巷”“五堡”“五祠堂”，总建筑面积达20万平方米以上。院外有院墙环绕，各院落既相对独立，又被统一在全院之中。院内采用大量砖雕、木雕进行装饰，是清代雕刻艺术中的精华。

The courtyard door of the Wang's Grand Courtyard in Shanxi Province

The Wang's Grand Courtyard was built by Wang's clan since the period during the reign of Kangxi in the Qing Dynasty. Through continual expansion, it finally became a large-scale house containing five alleys, five castles, and five ancestral halls and with a total floorage of over 200,000 square meters. The compound is encircled with walls and the courtyards inside are separated from each other. Decorated with a large number of brick carvings and wood carvings, the compound is the essence of the art of carving in the Qing Dynasty (1616-1911).

北京民居

北方民居以北京民居为代表，其中最具特色的是北京四合院。四合院是中国历史上最悠久、应用范围最广的民居形式。所谓四合，“四”指东、西、南、北四面，“合”即四面房屋围在一起，形成一个“口”字形的结构，里面是一个中心庭院，所以这种院落式民居被称为四合院。北京四合院的建筑结构为抬梁式木构架，梁柱为承重

Beijing Folk House

Beijing folk house is the representative of the folk houses in North China. The most typical of it is Beijing courtyard house, which is the oldest and the most widely used folk house. As its name suggests, the courtyard house has a central courtyard surrounded by houses on all four sides. Beijing courtyard house is in wooden post and lintel construction. Its beams and columns form the weight-carrying structure and its doors, windows, and partition boards are all made of

- **老舍故居的七彩木影壁**

中国著名作家老舍先生晚年在这座北京四合院里生活了16年。

The Multicolor Wooden Screen Wall of the Old Residence of Lao She, a Famous Chinese writer

Mr. Lao She lived in this courtyard house in Beijing for 16 years, where he spent the last years in his life.

结构，门窗、隔扇等也均为木制，周围则以砖砌墙。梁柱门窗等木构件一般都会施以油漆彩画，虽然没有宫廷园林建筑的金碧辉煌，但也色彩缤纷。

wood. Surrounding the house are brick walls. All the wooden components such as beams, columns, doors, and windows are painted and decorated. Though not as resplendent and magnificent as imperial garden, they are no less colorful.

影壁

影壁，又称“照壁”，是中国传统建筑中用于遮挡视线的墙壁，即使大门敞开，外人也看不到院内。

Screen Wall

The screen wall is adopted in traditional Chinese buildings to block the view of the outsiders. Even if the gate is wide open, outsiders cannot see into the courtyard.

• 北京四合院的院门
The Gate of a Beijing Courtyard House

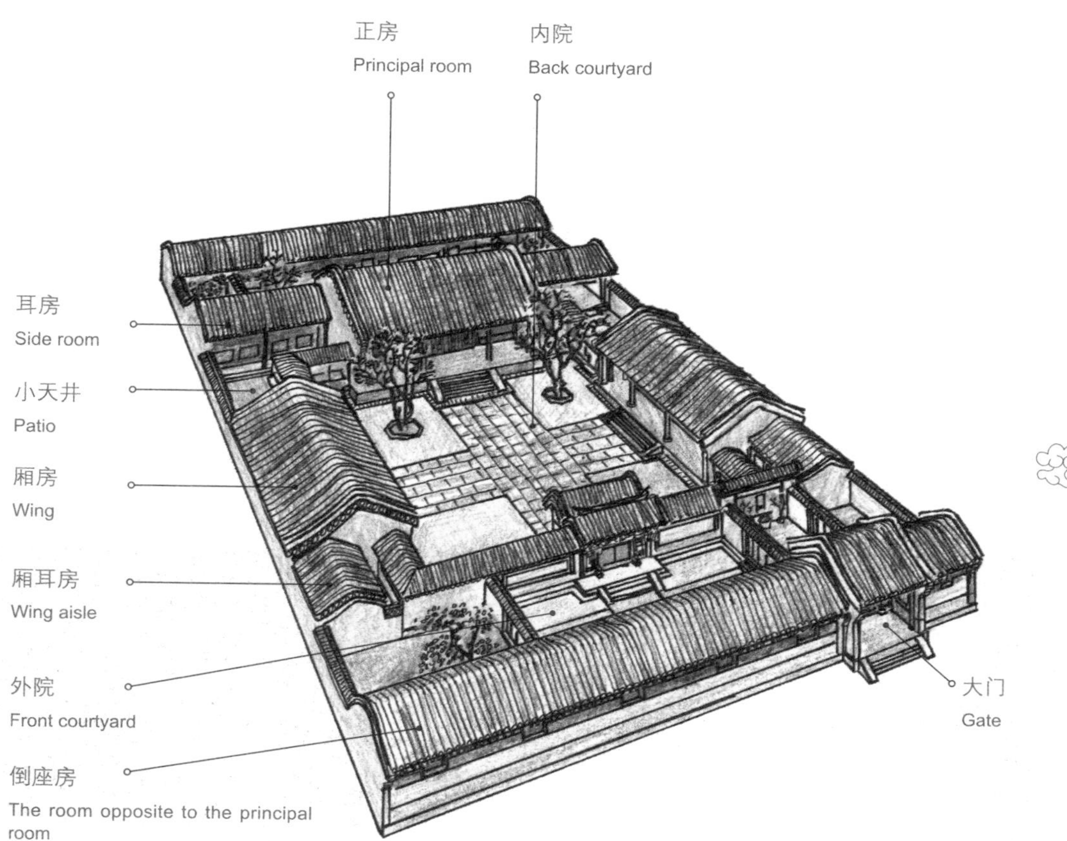

• 北京四合院结构图
Structure Drawing of a Beijing Courtyard House

客家民居

客家人是汉族的一个支系，最具代表性的客家民居建筑为土楼，其最大的特点就是防御性强。客家土楼一共46座，由六群四楼组成。土楼如果按形状来分类，可分为圆楼、方楼、五凤楼，此外还有变形的凹字形、半圆形、八卦形等等，其中以圆楼与方楼最为常见。土楼一般高三至五层，一层是厨房，二层是仓库，三层以上是起居室，一座土楼大约可居住200～700人。土

Hakka Folk House

Hakka is a branch of Han people. Their most representative folk house is called Tulou, which guarantees a strong defence. Now, there are 46 Hakka Tulou and they are four-storey buildings in six groups. The shape of Tulou can be divided into round, square, and Five-phoenix. There are also other shapes such as concave, half-round, and the Eight Diagrams. The round shape and square shape are the most common. Tulou is normally three to five-storey high. The ground floor serves

• **客家土楼群**

客家人建筑的土楼，可以称得上是全世界独一无二的大型民居。这些土楼造型独特、规模宏大，所体现的客家人的聚居方式以及土楼与周围环境的融合，都令人赞叹。

Hakka Tulou Group

Hakka Tulou are indeed the only large-scale folk houses in the world. They merge perfectly into the surrounding environment. Their unique shape and large size indicate the living habit of Hakka people. All these are the marvelous features of Hakka Tulou.

• **集庆楼内景**

集庆楼每户住家门前都立有一架木梯，这样一来就保证了各户人家的私密性，使每户都有独立的空间。

An Interior View of the Jiqing Tulou

In the Jiqing Tulou, each family has a separate wooden ladder to get in and out of their home. Such arrangement guarantees privacy and a separate space for each family.

楼以木结构为主，木结构组成土楼房间的构架，门、窗、走廊、围栏都用木材制成。裸露在外面的木材，很少有木雕等工艺装饰，所以其木料的天然纹路和造型得以显现出来。

as kitchen, the second floor as warehouse, and the third and upper floors as living rooms. A single Tulou can accommodate 200 to 700 people. Tulou is mainly in wooden structure as its room frame, door, window, corridor, and enclosure are all made of wood. The exposed wood is seldom decorated with carvings and its natural grain and shape can be easily seen.

承启楼内部的顶架结构

承启楼内部的木制顶架结构，从上至下递减，这样可以保护木结构不受雨水的冲刷。

The Interior Upper-frame Structure of the Chengqi Tulou

This is the interior wooden upper-frame structure of the Chengqi Tulou. With its roof layers shortened from above down, its wooden structure is protected from rain.

土楼内部的木结构

Wooden Structure inside Tulou

土墙：高大结实的土墙增强了土楼的防卫功能，圆形土楼一二层对外不开窗，其安全性就大大提高了。

Earth wall: The high and sturdy earth wall greatly enhances the building's defence. With no windows on the first and second floors, the building is more secured.

望台：集庆楼高大的墙面上，设立了望台，人们可立于其上观望村口的动向。

Lookout platform: Lookout platforms are set on the high walls of the Jiqing Tulou. From the platform, the residents can observe the situation at village entrance.

起居室：最外圈的环形建筑是人们日常起居的地方。

Living room: Living rooms are inside the outer ring of the building.

祖堂：祖堂是客家人举行祭祀、议事、庆典等重大活动的场所，也是土楼内的核心建筑，所以都是建在土楼的中心位置，祖堂的大门正对着正大门。

Ancestral hall: The ancestral hall is the venue for Hakka people to hold important events such as sacrifice giving, meeting, and celebration. As the core structure inside the Tulou, it is always in the center. The door of the ancestral hall directly faces the front gate of the Tulou.

- **集庆楼结构示意图**

集庆楼建于明朝，坐南朝北，占地面积达2826多平方米，为抬梁穿斗混合式构架。集庆楼外环为四层，每层有56个开间。最初集庆楼为内通廊式，后来两层以上改为单元式，每个单元各有一道楼梯，单元之间的廊道有木板相隔。

The Structure Sketch Map of the Jiqing Tulou

Built in the Ming Dynasty (1368-1644) and taking up an area of over 2,826 square meters, Jiqing Tulou sits on the south and faces the north. It is in a style combining the post and lintel construction with the column and tie construction. Its outer ring has four storeies and each contains 56 rooms. Jiqing Tulou initially had an inner corridor. Later, it was changed into the unit style from the second floor up. Each unit has a stairway and the corridor between the units is blocked with planks.

吊脚楼

有一些少数民族的民居也为木结构，其中较为典型的是吊脚楼。吊脚楼在中国的湖南西部、贵州等地常见，也是苗族、壮族、布依族、侗族、土家族等少数民族的传统民居形式。吊脚楼大多依山就势而建，一般为东西向，底层用来堆放物品，二楼住人。二楼设有厅，

Stilted House

Some China's ethnic groups also build their folk houses with wood. The elevated house is a typical one. Commonly seen in the west of Hunan Province and Guizhou Province. It is a traditional folk house of Miao ethnic group, Zhuang ethnic group, Buyi ethnic group, Dong ethnic group, and Tujia ethnic group. Most elevated houses are built at the foot of a mountain and their windows usually face the east and west. The ground floor is for storing articles and the second for living. There is a parlor on the second floor for receiving

• 湖南省湘西吊脚楼
Stilted Houses in West Hunan Province

用来接待客人，三层的吊脚楼除设有起居室外，还有隔出来的小间用来储存粮食或物品。除了屋顶盖瓦以外，吊脚楼全用杉木建造而成。屋子的杉木柱上面凿有洞，用大小不一的小杉木将柱与柱套连起来，建筑虽不用铁钉，却非常牢固。

guests. On the third floor, the last one, are living room and small apartments for storing food or articles. Except the tiled roof, the entire house is built with fir wood. Holes are chiseled into the main columns and different sizes of fir wood tie beams are used to bind the columns together. Though without a single nail, the building stands firmly and steadily.

侗寨鼓楼

鼓楼是贵州和广西壮族自治区的侗族村寨中一种极富特色的民族建筑，也是侗族祭祀祖先神的地方。鼓楼是侗族村寨中的公共建筑，一般建于寨子中心，是全寨集会、议事和文化娱乐的场所。鼓楼的结构形式有宝塔、宫殿、方亭等几种类型。侗寨鼓楼底层呈方形，内部的大厅可容纳数百人。上部重檐相叠，有的高达十二三层。檐下绘有具有民族特色的图案。侗寨鼓楼用杉木建造，梁枋、柱子等构件纵横交错，不施一钉，但坚实牢固。顶楼常悬一面鼓，如果遇到大事就会击鼓召集众人集会。

Drum Tower of Dong People

The drum tower is a characteristic building of the Dong people living in Guizhou Province and Guangxi Zhuang Autonomous Region. It is the venue for Dong people to offer sacrifice to their ancestors and gods. A drum tower is a public building in a Dong Village. It is normally built at the center of the village. Dong people also gather at the tower to have meetings, discuss village affairs, and carry out various cultural and recreational activities. The tower can take the shape of a pagoda, palace, or square pavilion. Its square ground floor is a hall that can accommodate several hundred people. Its upper part has overlapping eaves, which sometimes can be as many as twelve or thirteen layers. Under the eaves are patterns with characteristics of Dong ethnic group. The drum tower of Dong people is built with fir wood. Its components such as beams and columns are crisscrossed to form an integral whole. Without a single nail, it is robust all the same. A drum is often hung on top of the tower. When there is a big event, villagers will be summoned by the roll of the drum.

• 贵州黎平肇兴侗寨

Zhaoxing Dong Village in Liping of Guizhou Province

• 贵州黎平肇兴侗寨鼓楼

The Drum Tower of Zhaoxing Dong Village in Liping of Guizhou Province

杉木

杉树高大通直，体量较大，纹理松且平直，木质较软，不开裂，耐腐性强，有天然的原木香味。杉木主要用于房屋、桥梁、造船、门窗、器具、造纸、纺织等。中国古代工匠常用杉木做家具、门、窗等，制式美观，自然通直的纹理与门、窗的平直相得益彰。

Fir Wood

Fir tree is tall and straight, hence the large fir wood. The fir wood has loose and straight grains, a relatively soft texture, and does not crack. It performs well in resisting corrosion and has the natural fragrance of the log. Fir wood is mainly used to build house and bridge, and to make boat, door, window, utensil, paper, and instrument for textile industry. The carpenters in ancient China used fir wood to make beautiful furniture, door, and window. The straight wood grain adds beauty to the upright doors and windows.

• 杉木佛龛

A Fir Wood Niche

> 其他木建筑

钟鼓楼

中国古代城市都建有钟楼和鼓楼，简称“钟鼓楼”。因为古时候没有机械钟表，确定每日时刻很困难，所以便由专门的官员以击鼓方

> Other Wooden Buildings

Bell and Drum Towers

All ancient cities of China have bell and drum towers. Without mechanical clock or watch, the ancient people could hardly determine the exact time. The government thus assigned an official to report the time by beating the drum. For example, the official would beat the drum five times a night, once every

• **西安钟楼**

西安钟楼是中国古代遗留下来的众多钟楼中形制最大、保存最完整的一座。钟楼建在方型基座上，基座之上为两层木结构楼体，总高36米，内有楼梯可盘旋而上。两层楼的四角均有木柱回廊、花窗及雕花门扇以及藻井、木雕、彩绘等装饰，是一座具有浓郁民族特色的宏伟建筑。

The Xi'an Bell Tower

Among China's numerous ancient bell towers still standing today, the Xi'an Bell Tower is the largest and the best-preserved. It is a two-storey wooden structure built on a square foundation. Some 36 meters tall, it has an internal spiral staircase. At the four corners on both floors are wooden pillars and corridors decorated with carved doors, windows, caisson ceiling, wood carving, and colored painting. It is a majestic building with the characteristics of the Chinese nation.

式报时，作为当地的标准时。比如，晚上报时的鼓声叫“更鼓”，一夜要敲击五次，一更即一鼓，三更即三鼓。修建了鼓楼，全城的人就都能听到鼓声。此外，中国古代有宵禁制度，也就是到了晚上，城门会关闭，禁止百姓出入或随意走动。比如唐代，长安城每到晚上，钟鼓敲击后，城门就会紧闭，街上禁止通行，除非是产妇或病人等。

two hours. A tower was built to house the drum so that all city residents could hear the drum roll. In addition, curfew was a regular practice in ancient China. In the Tang Dynasty (618-907), for example, Chang'an City would be under curfew after the drum was beaten at the beginning of the night. The city gate was closed and no one was allowed to pass the check points in the street, except the delivery of lying-in women or patients with valid certificate.

• 北京鼓楼

北京鼓楼位于城市的中轴线上，梁枋立柱均为木制。

The Beijing Drum Tower

The Beijing Drum Tower stands on the axle line of the city. Its beams and columns are all made of wood.

木牌坊

牌坊，又名“牌楼”，是中国封建社会为了表彰功勋及忠孝节义所立的建筑物，还有一些宫观寺庙以牌坊作为山门，因此多建于宫苑、寺观、陵墓、祠堂、衙署和街道路口等地方。牌坊最早只是一种两柱式的木建筑，后来发展成在柱子上加屋顶，也就是牌楼。牌楼一般较为高大，主要由木、砖石、琉璃几种材料建成。古代木牌楼数量最多，一般用柏木做桩，每根柱子的下部用“夹杆石”包

Wooden Archway

Also known as gateway, the arch was the structure erected in feudal China to honor great feats and the acts of loyalty, filial piety, chastity, and righteousness. Some palaces and temples also had arches as their gate at the foot of the mountain. Therefore, arches were mostly built at palaces, temples, mausoleums, ancestral halls, government offices, and street crossings. The earliest archway was a two-column wooden structure. Later, a roof was mounted on the top to form the

- **颐和园苏州街**

颐和园中的苏州街全长300多米，商铺林立，共建有牌楼19座。

The Suzhou Street in the Summer Palace

The Suzhou Street in the Summer Palace is over 300 meters long. Lined on both sides with countless shops, it has 19 gateways in total.

• 山西平遥古城“千祥云集”牌楼

The Gateway of Auspicious Clouds Gathering in the Old Town of Pingyao, Shanxi Province

• 北京北海“积翠”牌楼

这座牌楼主要用木材建成，顶部设有楼顶、匾额等装饰。

The Gateway of Green Accumulation in Beihai Park, Beijing

Built mainly with wood, this gateway has roofs and inscribed boards on the top as decorations.

住，外面再束以铁箍。柱顶覆瓦，防止风雨侵蚀木柱。

清代，木牌楼被广泛应用于商业街，成为大商铺门前必不可少的建筑。当时，木牌楼在民间也称为“风水柱”，柱子越高寓意越能发财。

木桥

木桥是中国最早出现的桥梁形式，它重量轻，加工建造容易。古代人将原木柱打入河床中作为桥

• 江西的彩虹桥

彩虹桥建于宋代，是廊桥的典型代表。

The Rainbow Bridge of Jiangxi Province

The Rainbow Bridge was built in the Song Dynasty (960-1279). It is a typical representative of the covered bridges.

gateway. A gateway was relatively large and tall and was built with wood, brick, stone, and colored glaze. Of all ancient gateways, wooden ones were the most common and they usually had cypress columns, each being wrapped up at the bottom by stones held together with iron hoops. Tiles were laid on the top of the columns to prevent the wooden columns from corroding by wind or rain.

In the Qing Dynasty (1616-1911), wooden gateways were built in business streets as an indispensable structure in front of all large stores. At that time, wooden gateways were called geomantic pillars. The higher the pillar, the more wealth it would bring.

Wooden Bridge

The earliest bridges in China were made of wood due to its light weight and easy-to-process property. The ancient people drove logs into river bed as bridge piers and laid planks on them. Though easy to build, wooden bridges are also easy to burn and rot. They can neither support heavy weight nor last

墩，然后在桥墩上铺木板，方便建造。但木桥易燃、易腐蚀，承载力和耐久性都不强，因此保存下来的很少。秦汉时期，木桥普遍改用石桥墩，以木材为梁。后来，人们又在木桥上修建了保护桥身的桥屋，于是出现了廊桥。

long and hence few have preserved. In the Qin and Han dynasties, most wooden bridges were replaced by those with stone piers and wooden beams. Later, people built bridge house to protect the bridge body and the covered bridge came into being.

• 云南丽江古城中古老的木桥
Ancient Wooden Bridges in the Old Town of Lijiang, Yunnan Province

柏木

柏木呈黄褐色，木质坚韧、细腻，抚之如幼童肌肤，耐腐朽，还有芳香的气味，属于较为名贵的木材。中国古代帝王将生长了数百年的香柏木封为“将军树”，在一些地方更被众人冠以“神木”的美名。中国栽培柏木历史悠久，常见于庙宇陵园，在园林、寺庙、名胜古迹处，常可以看到古柏参天。此外，中国古代军队多用柏木来制作弓箭，后来开始用于建筑、造船、装饰、器具等，上好的棺木也用柏木。

• 柏木
Cypress Wood

Cypress Wood

Cypress wood has a yellowish brown color and a fine, smooth, and tenacious texture. It touches like the skin of a young kid and performs well in resisting rot. Sending forth sweet fragrance, it is a rather precious wood. The emperors in ancient China conferred the title of general on the several-century-old cypress trees. In some places, cypress trees are even praised as the divine tree. Cypress planting has a long history in China and towering cypress trees are commonly seen in temples, mausoleums, gardens, and places of historic interest and scenic beauty. Moreover, cypress wood was often used to make bows and arrows for the army in ancient China. Later, it was used to build houses and boats, make utensils, and apply decorations. The best cypress wood was also used to make coffins.

栈道

栈道，又称“阁道”“复道”，是沿悬崖峭壁修建的一种道路。栈道一般是在地势险要的悬崖峭壁上凿石孔，然后在石孔中插上石桩或木桩，上面横铺木板或石板，可供行人和通车。秦汉时期，在巴蜀地区曾修建过长达千里的古栈道，后人

Plank Road

The plank road has several other names in Chinese. People built it on cliff face by inserting stone or wooden piles into the holes chiseled on the cliff and paving planks or slabs on the piles. Plank road is built for both pedestrians and carts. In the Qin and Han dynasties, people once built thousand-mile plank road in Sichuan

• 山上的木栈道（图片提供：全景正片）
A Wooden Plank Road on Mountainside

均有维修。现代栈道多是园林里富有情趣的楼梯状木质道路，被称为“木栈道”。

Province, which was repaired repeatedly in the following dynasties. Modern plank roads are normally the wooden ones built as stairway in gardens for fun. They are called the wooden plank road.

索桥

索桥，又称“吊桥”，是用藤、铁索等为骨干，上铺木板，悬吊起的大桥，一般建在水流湍急、不易建桥墩的陡岸险谷。过索桥时，人会感觉十分惊险，整个身体悬在半空中，桥身晃动，仿佛随时都会掉下去。建造索桥先要在两岸建屋，屋内设立柱和转轴，分别用以系绳和稳绳，然后把若干根粗绳平铺系紧，再在绳索上横铺木板，有的索桥还会在两侧加一两根绳索作为扶栏。

Chain Bridge

The chain bridge, or suspension bridge, is the bridge hung on vines or iron chains and paved with planks. They are normally built on steep banks and precipitous valleys where the water runs fast and bridge piers can hardly be built. People will feel the thrill when crossing a chain bridge. Hung in midair, they walk on the bridge, sway with it, and have the feeling of falling at any moment. To build a chain bridge, a shed should be built on either bank and the upright posts and revolving shafts be set in the sheds for tying and stabilizing the chains. Then, several thick ropes are tied around the chains and planks are laid on the ropes. Some chain bridges have one or two side ropes as handrails.

• 索桥
A Chain Bridge

工艺精湛的木家具

Superbly-crafted Wooden Furniture

中国的木家具历史源远流长，式样由低矮到高大，由单一到多样，积淀了深厚的文化内涵，具有强烈的民族风格。尤其是明清时期，古典家具更是发展到了一个巅峰。明式家具造型简洁典雅，做工精妙，在实用与美观相统一中洋溢着古朴雅致的美。清式家具吸收了外来文化，形成了造型浑厚、讲究富丽繁缛的装饰美。中国古典家具可分为床榻类、椅凳类、桌案类、箱柜类、屏风类和架具类等，皆各具特色。

Chinese wooden furniture has a long history. Developing from short to high and from unitary to diversified, Chinese wooden furniture has taken with profound cultural connotation and rich national characteristics. The classical furniture reached a new height in the Ming and Qing dynasties. Furniture in the Ming Dynasty shows a simple and elegant style and exquisite workmanship. It demonstrates a graceful beauty created by the combination of its practical and aesthetic values. Furniture in the Qing Dynasty absorbs foreign cultures and forms a decorative beauty stemmed from its vigorous appearance and gorgeous design. Chinese classical furniture can be divided into many varieties, including bed and couch, chair and stool, table and desk, case and cabinet, screen, and shelf. They all have their own characteristics.

> 床榻

床和榻都是供人睡卧休息的家具，因此常被并提。其实，床榻在中国的历史非常悠久，早在远古时期，人们就用树叶、茅草或兽皮做成“地铺”，后来逐渐用木材制作床榻。

床

床是供人躺在上面睡觉的家具。中国的床早在春秋战国时期就已经出现了。发展到明清时期，床的形制高大起来，宛如一间雕梁画栋的房子，强调密闭性。这时的床一般较宽大，能睡双人，摆放在居室中的暗间，有架子床、宝座床（龙凤床）、拔步床等式样。清式床在康熙以前保留着明代的特点。到了乾隆时期，形成了用

> Bed and Couch

The bed and couch are both the furniture for people to sleep and rest on. Therefore, they are mentioned as a whole by people. In fact, bed has a very long history in China. In ancient times, people made beds with leaves, thatch or hides. Later, they gradually used wood to make beds.

Bed

The bed is the furniture for people to sleep on. In China, the earliest bed appeared as early as the Spring and Autumn Period and Warring States Period. By the Ming and Qing dynasties, the bed became so big and enclosed that it looked like a house with carved beams and painted pillars. The beds at that time were wide enough for two people and they were put in the dark

围栏：明清时期的床多设围栏，上面多雕有各种花纹，非常精致。

Railing: Most beds made in the Ming and Qing dynasties have surrounding railings carved with many exquisite patterns.

百子图图案：百子图又叫“百子迎福图”“百子嬉春图”，源于周文王（中国古代的一位君王）生百子的典故，是祥瑞之兆。画面常画众多小孩，寓意多子多孙，多福多寿。

The Painting of One Hundred Sons: The Painting of One Hundred Sons is also called the Painting of One Hundred Sons Greeting Fortune and Painting of One Hundred Sons Having Fun in Spring. According to the legend, Emperor Wen of the Zhou Dynasty gave birth to one hundred sons, a propitious sign for good fortune and longevity.

床屉：架子床的床屉一般分为两层，上层用绳和藤编织而成，或用木板制成。

Bed tray: A canopy bed usually has two layers of bed trays. The upper layer is woven with ropes and vines or made with planks.

清代楠木架子床

架子床是有柱子承托床顶的双人床的统称，有多种形制，便于悬挂蚊帐和锦帐。

A Phoebe Zhennan Canopy Bed of the Qing Dynasty (1616-1911)

Canopy bed is the general term for the double bed with a top supported by pillars. There are different types of canopy bed, all designed to facilitate the hanging of mosquito net and brocade screen.

材厚重、装饰华丽的清式风格，家具制作力求繁缛多致，不惜耗费工时和木材。

红木

红木材质坚硬耐用，有光滑细密的纹理，产量较大，因此是紫檀木、黄花梨木等上等木材日渐匮乏之后的替代品。用红木制作家具时常选取红木材中最精美的部分，因此较为名贵。红木家具纹理自然，平整润滑，光泽耐久，给人一种淳厚含蓄的美感。

Mahogany

Mahogany is hard and endurable. It has smooth and fine grains. Produced in large quantity, it is a good substitute for the top-class wood varieties that are getting scarcer and scarcer, including red sandalwood and yellow rosewood. The finest part of mahogany is chosen to make furniture, hence its high value. Mahogany furniture has natural grain, an even and smooth touch, and endurable gloss, presenting a pure and implied beauty.

• 红木切面
A Section of Mahogany

corner of the bed room. There were many styles, including the canopy bed, the throne bed (dragon and phoenix bed) and the alcove bed. Ming beds remained popular in the Qing Dynasty until the reign of Kangxi. By the period during the reign of Qianlong, Qing-style bed had taken shape in heavy materials and brilliant decoration. The carpenters at that time tried their best to make the bed as sophisticated and intricate as possible, regardless of the time and materials spent in the process.

床顶：清代的拔步床床顶安盖，刻有精美的吉祥图案，雕饰繁缛。

Bed top: The alcove bed of the Qing Dynasty (1616-1911) has a capped bed top, which bears exquisite auspicious patterns and sophisticated carvings.

窗户：窗户透雕精美。

Window: The window has beautiful fretwork.

踏步：长出床沿1米左右的平台，在床前形成一个小廊。

Step: It is a platform one meter out of the bed edge, forming a small corridor in the front of the bed.

底座：一个木制平台，床置于其上。

Pedestal: It is a wooden platform on which the bed is placed.

木围栏：多为三面围栏，上面雕刻有精美的吉祥图案。

Wooden railing: There is usually a three-sided railing carved with exquisite auspicious patterns.

- **清代拔步床**

拔步床形制庞大，从外观看好像是一个木屋。拔步床由两部分组成：一是架子床；二是架子床前的围廊，围廊与架子床相连为一个整体。床前有廊，有相对独立的活动范围，廊的两侧可以放置桌凳、便桶、灯盏等小型家具，跨入廊就好像进入了室内。

An Alcove Bed of the Qing Dynasty (1616-1911)

The alcove bed is huge and looks like a log cabin. It is composed of two parts: the canopy bed and the surrounding railing, which are connected into an integral part. The railing forms a relatively separate space. On both sides of the railing, small furniture such as stool, chamber pot, and lamp can be placed. Stepping into the railing is like entering a room.

榻

榻是一种狭长而低的坐卧用具，最早出现在西汉后期，专供一人坐卧。发展到明清时期，榻大都陈设在厅堂中，大多仅容一人坐卧，又名“独睡”，有罗汉床、宝座、贵妃榻、床柜等形式，一般陈设在正房明间，供主人休息和接待客人之用，其功能相当于现在的双人沙发。

Couch

The couch is a long, narrow, and low-lying furniture for people to sit on or lie down. The earliest couch, which was for one person, appeared in the late Western Han Dynasty. By the Ming and Qing dynasties, couches were normally placed in the hall and they remained the one-person furniture. Dubbed the Single Sleep, there were various types of couches, including the Luohan bed, throne, beauty's couch, and bed chest. They were normally placed in the bigger outer room of the main house for the house owner to rest on or receive guests. They served as the double sofa at present.

近代贵妃榻

贵妃榻又叫“美人榻”，是古时妇女用来小憩的一种形制狭小、可以坐卧的榻，造型优美，制作精致。

A Beauty's Couch Made in Modern Times

The beauty's couch was made in ancient times for women to sit on or lie down. Though narrow, it was beautifully designed and exquisitely made.

暗八仙纹：一种宗教纹样，具有祝颂长寿、驱邪保平安的寓意。暗八仙纹中以八仙（八仙：民间广为流传的道教八位神仙，即汉钟离、吕洞宾、蓝采和、韩湘子、铁拐李、曹国舅、张果老、何仙姑）所持之物代表各位神仙，包括汉钟离的芭蕉扇、吕洞宾的宝剑、蓝采和的花篮、韩湘子的笛子、铁拐李的宝葫芦、曹国舅的阴阳板、张果老的渔鼓、何仙姑的莲花或荷叶。

Hidden grain of eight Taoist immortals: It is a religious pattern for celebrating longevity and expelling evil spirit. The hidden grain shows the signature articles used by The Eight Immortals (Eight well-known Chinese mythological immortals of Taoism: Han Zhongli, Lv Dongbin, Lan Caihe, Han Xiangzi, Li Tieguai, Cao Guojiu, Zhang Guolao, He Xiangu) to represent them. They are the palm-leaf fan of Han Zhongli, the sword of Lv Dongbin, the flower bucket of Lan Caihe, the flute of Han Xiangzi, the magic calabash of Li Tieguai, the Yin-Yang board of Cao Guojiu, the bamboo drum of Zhang Guolao and the lotus or lotus leaf of He Xiangu.

嵌大理石：围屏中镶嵌着带有山水花纹的大理石，构成了一幅天然的水墨山水画。大理石一般产自云南，有美丽的纹理。

Marble inlay: The screen board is mounted with the marble bearing the landscape grain, making the Luohan bed a natural Chinese ink and wash painting of landscape. Marble is generally produced in Yunnan Province and has beautiful grains on it.

清代红木罗汉床

罗汉床是左右两侧和后面装有屏板，但不带立柱、顶架的一种床榻，长约2米，宽约1米，高约40厘米，能用于坐卧。

A Mahogany Luohan Bed of the Qing Dynasty (1616-1911)

The Luohan bed has screen boards on its left, right, and back sides but has no upright post or top stand. The bed is about two meters long, one meter wide, and 40 centimeters tall. It is for people to sit and sleep on.

> 椅凳

椅凳类家具包括椅、凳、墩、杌等坐具。在椅凳类家具出现之前，席和床榻类家具一直是中国古代人最主要的坐具。直到唐代中期以后，椅凳类家具才进入上流社会，应用于宫廷宴饮、家居生活及行军战场。

椅

椅是有靠背的高型坐具，品种多样，有些还有扶手，现在所见的椅子大多是明清以来的椅子式样。从明代起，椅的式样多了起来，而且把精巧、实用的传统美学观念和

> Chair and Stool

The chair-like furniture includes chair, stool, drum stool, and folding stool. Before their advent, seat, bed, and couch were the main furniture for the ancient Chinese to sit on. Chairs and stools were not adopted by the high society until after

• 鹿角椅
An Armchair Made of Deer Horns

人体结构结合起来，创造了风格简约、舒展大方的明式家具。特别是明代中晚期，出现了用黄花梨木、紫檀木、铁力木、酸枝木制作的明式硬木椅子。清代椅在明式的基础上又有所变化，主要是增加了雕刻装饰，变肃穆为流畅，化简素为雍容，并加大了椅子的尺寸。

• **明代矮腿圈椅**

圈椅是扶手椅的一种变体形式，是将靠背与扶手连成圈形，靠背板一般向后弯曲，中心多有装饰图案。

A Short-leg Round-backed Armchair of the Ming Dynasty (1368-1644)

The round-backed armchair has its back connected with its armrest. Its back board curves backward and there are normally decorative patterns on its center.

mid Tang Dynasty. Since then, it had been used in court banquets, daily life, and barracks.

Chair

The chair is a high seat with a back. Chairs are in diversified styles and many of them are exquisitely made. Some chairs even have arms. Most of the present-day chairs still take the styles designed in the Ming and Qing dynasties. Since the Ming Dynasty, chairs have been made in more and more styles. Ming-style chairs are in simple and unaffected style. They combine the practical traditional aesthetic values with the structure of human body. In mid and late Ming Dynasty, the hardwood chairs made of yellow rosewood, red sandalwood, ironwood, and blackwood emerged. Qing-style chairs had further changes based on the Ming style. The main change was the addition of carving decorations, which turned solemnity into smoothness, and simplicity into elegance. In addition, Qing-style chairs became larger.

透雕装饰：透雕是明清家具的三大主要雕刻手法（浮雕、透雕和圆雕）之一。先设计好纹样，然后将多余的部分全部挖掉，镂空突出纹样，将留出的图案做成立体效果。

Fretwork decoration: Fretwork is one of the three major carving techniques used for furniture in Ming and Qing dynasties. The other two are relief sculpture and circular engraving. First, the pattern is designed. Then, the unwanted parts are dug out until the pattern stands out. Finally, the pattern is processed to show a 3D effect.

- **清代花梨木玫瑰椅**

玫瑰椅是为坐于书桌旁写作而设计的，大多用花梨木或鸡翅木制作，靠背和扶手与椅座均为垂直相交，同时靠背较低，与扶手高低相差不大。

A Rosewood Rose Chair of the Qing Dynasty (1616-1911)

The rose chair is designed for writing practice at a desk. It is usually made of rosewood or chicken-wing wood. Its back and armrest are both vertical to the seat. The back is lower to the height of the armrest.

明代灯挂椅

灯挂椅的靠背体窄而高，呈“S”形，适合人体背部曲线，外形很像江南农村使用的油盏灯的提梁，故而得名。灯挂椅整体多通光无雕饰，有装饰的灯挂椅也仅在靠背上雕一简练精美的图案。黄花梨、红木、榉木、铁力木由于纹路很清晰，木质坚硬，是明清灯挂椅的主要材料。

A Lamp-hanger Chair of the Ming Dynasty (1368-1644)

The lamp-hanger chair has a narrow, high, and S-shaped back that fits in well with the curve of human back. It derives its name from its likeness to the hanger of the oil lamp commonly used in the rural areas south of the Yangtze River. The chair is normally smooth without carving decoration. Even the decorated one bears only one simple and beautiful pattern on its back. Yellow rosewood, mahogany, beech wood, and ironwood have clear grains and hard textures. They are the main material for making the lamp-hanger chair in the Ming and Qing dynasties.

花梨木

花梨木色彩鲜艳，纹理细而匀称、清晰美观，木质硬细、沉重且强度高，耐久耐腐性强，与黄花梨木非常相似，常用来制作家具及文房用品。在明清时期，花梨木家具非常盛行，清代有许多红木家具都是用花梨木制造的。

Rosewood

Rosewood has bright colors, a hard texture, and its grains are fine, even, clear, and beautiful. It is heavy and strong. It lasts long and performs well in resisting corrosion. It is very similar to yellow rosewood. It is often used to make furniture and stationery articles. In the Ming and Qing dynasties, rosewood furniture was popular. In the Qing Dynasty, many pieces of mahogany furniture were made of rosewood.

花梨木切面

A Section of Rosewood

榉木

榉木木质重且坚固、有光泽，纹理直且清晰，木材质地均匀，常带有美丽的大花纹，以塔型纹最佳，俗称“宝塔纹”。榉木家具多为大料，极为坚固耐用，纹理优美精致，制作工艺相当考究，用其制成的案、桌、柜门一般都是独板。明末清初，榉木家具追仿黄花梨木家具的神韵；清中叶以后，榉木家具又开始追仿商品化倾向十足的红木家具。

Beechwood

Beechwood is heavy, strong, and glossy. It has straight and clear grains, an even texture, and large beautiful decorative patterns. The pagoda-shaped pattern, known as pagoda pattern, is the best. Beech furniture is usually made of large pieces of wood, which results in its high firmness and high practical and aesthetic values. It is a work of art that demonstrates fine workmanship. Normally, beech desks, tables, and cabinet doors are made of solid board. At the turn from the Ming Dynasty (1368-1644)to the Qing Dynasty (1616-1911), beech furniture began to imitate the romantic charm of yellow rosewood furniture. Since mid Qing Dynasty, it began to imitate mahogany furniture, which had a great inclination towards commercialization.

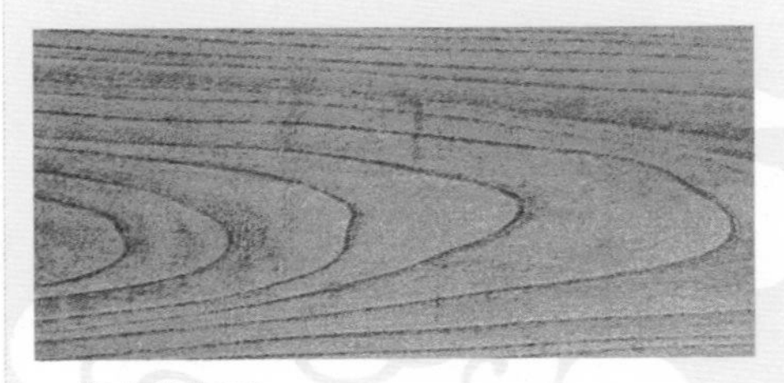

• 榉木的切面

A Section of Beechwood

• 清代红木灵芝太师椅

明清时期的太师椅没有固定式样，形体偏大，造型厚重庄严，成排陈设在厅堂之上。太师椅的上部和下部为独立的两个部分，上部是方形屏风式的靠背和扶手，垂直于椅面；下部为独立的杌凳，凳腿截面多为坚硬的方形，整体感觉雄壮厚重。

A Mahogany Ganoderma Master's Chair of the Qing Dynasty (1616-1911)

In the Ming and Qing dynasties, the master's chairs had no fixed style. They were normally large with a stately and solemn design and were often lined up in the hall. A master's chair has two separate parts: the upper part and the lower part. The upper part includes the back and armrest in the square screen style and is vertical to the seat. The lower part is a separate stool and the stool leg section is often a hard square. The chair as a whole looks strong and decorous.

清代楠木一统碑椅

一统碑椅无扶手，靠背外轮廓为规整的长方形，很像一座碑，故名。

A Phoebe Zhennan Stele Chair of the Qing Dynasty (1616-1911)

The stele chair has no armrest. Its back is an oblong with a neat contour. It looks like a stele, hence its name.

明代榉木官帽椅

官帽椅因其形似中国古代的官帽而得名，用材或圆或方，或曲或直，背板通常做成“S”形曲线，主要流行于中国南方。

A Beech Official's Hat Armchair of the Ming Dynasty (1368-1644)

The official's hat armchair derives its name from its likeness to an official's hat in ancient China. It is made of all shapes of wood, either round or square, and either curved or straight. Its back board is usually made into an S-shape. Such chair is popular in South China.

清代红木躺椅

躺椅是一种靠背很高、又可大角度向后斜仰，椅座较长、带扶手的椅式，因人可以仰躺在上面而得名。

A Mahogany Deck Chair of the Qing Dynasty (1616-1911)

The deck chair has a high back, long seat, and armrest. Its back can be put down backwards at a large angle. Its design allows its user to lie down.

紫檀木

紫檀木产量稀少，在明清时期备受皇家推崇，被大量用于家具制作、工艺品雕刻等，但存世的稀少。紫檀木沉静古朴、庄重大方，木质坚实细腻、稳定性强、韧性好、耐雕琢、色泽雅致、纹理美观，因而紫檀木家具给人一种沉稳牢靠的感觉，并经久耐用，流传数百年而不毁损，历久弥新。紫檀木的这些特性十分符合明清帝王期望江山永固、富贵长存的心理需求，因此受到皇室的宠爱。

Red Sandalwood

Red sandalwood has a fine texture and crisscrossing grains. Hard to find, it was highly valued by royal families in the Ming and Qing dynasties. It was once used to make great quantities of furniture and artistic carvings, but few have survived to this day. Red sandalwood has a quiet, unsophisticated, poised, and unaffected appearance. Being strong, smooth, stable, and tensile, it is a good material for carving. Red sandalwood also has graceful colors and beautiful grains. The furniture made of it guarantees a stable and reliable feeling and can last as long as several centuries. These properties of red sandalwood fit in well with the emperors' psychological demand for long-lasting rule and prosperity. It was naturally loved by royal families.

紫檀木原木

Red Sandalwood Logs

靠背：透雕麒麟纹。麒麟是中国古代传说中的一种瑞兽，与凤、龟、龙并称为“四灵”。麒麟纹是古代装饰中常见的装饰纹样，寓意吉祥，事业有成。

Back: The back has kylin pattern fretwork. Kylin is an auspicious animal in ancient Chinese legends. It is among the four divine animals. The other three are phoenix, tortoise, and dragon. Kylin pattern is common in ancient decoration. It represents good luck and successful career.

明代黄花梨木交椅

交椅是一种折叠式的椅子，有圆后背交椅与直后背交椅之分，用材以黄花梨木为贵。

A Yellow Rosewood Folding Chair of the Ming Dynasty (1368-1644)

The folding chair is foldable. There are two kinds of folding chairs, one with a round back and the other with a straight back. Those made of yellow rosewood are the most precious.

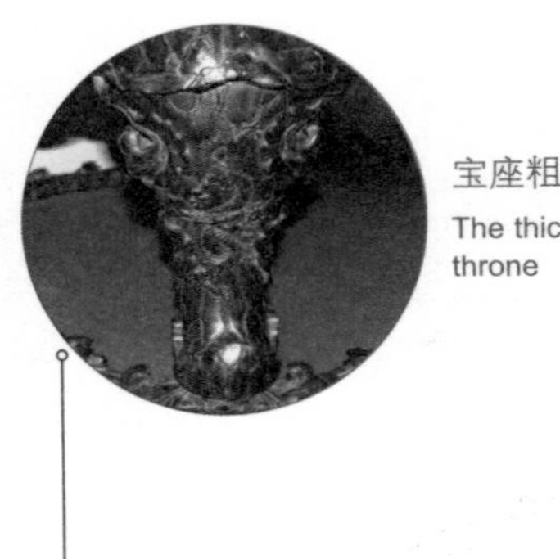

宝座粗壮的腿部

The thick leg of the throne

宝座上的浮雕龙纹，华贵精美。

The relief dragon patterns on the throne are luxurious and exquisite.

- **清代紫檀木雕荷叶龙纹宝座**

宝座又称“御座”，是中国古代宫廷中供皇帝日常生活使用的坐具，陈设在皇帝和后妃寝宫的正殿明间，象征皇权至高无上。宝座一般单独陈设，多用于隆重场合，如皇帝登基仪式、朝贺大典等，而且要放在最显著的位置。

A Red Sandalwood Throne of the Qing Dynasty (1616-1911) with Carved Lotus and Dragon Patterns

Also known as imperial seat, the throne was used by emperors in imperial daily life in ancient China. It was normally placed in the outer rooms of the main hall serving as the resting palace for the emperor and his concubines. Representing the supreme imperial power, the throne was normally placed separately and used mainly in solemn occasions such as the enthroning ceremony and grand celebrations. It should be placed in the most conspicuous position.

北京故宫太和殿的家具陈设

太和殿是故宫的主殿之一，重要大典和朝会都在这里举行。殿内有六根金漆蟠龙大柱，其他如金漆盘龙宝座、绚丽的彩画、蟠龙金凤藻井、铺地金砖等，组成了太和殿壮丽的布局形式。宝座和七重屏风等家具陈设在高近1.6米的地台上。宝座为贴金和上金漆装饰，通体金光熠熠，绚丽夺目，居中而设。宝座后设有七重屏风，装饰豪华，气势雄伟。地台设三路阶梯，每路有七级台阶，阶梯上铺大红地毯，中间的地毯上绣着巨型金龙云水图。

Furniture Layout in the Hall of Supreme Harmony of the Imperial Palace, Beijing

The Hall of Supreme Harmony is one of the main halls in the Imperial Palace. It was the venue for important ceremonies and court meetings. The hall houses many well-crafted items, including six columns coated with gold lacquer and twined round with dragons, a throne also with gold lacquer and dragon carvings, gorgeous painting decoration, a dragon and phoenix caisson ceiling, and gold bricks paved on the ground. They form the brilliant layout of the hall. The throne and the seven-panel screen are placed on a 1.6-meter-tall platform. The throne, decorated with gold leaves and gold lacquer and occupying the central position, is the shiniest object in the hall. The seven-panel screen behind the throne is luxuriously decorated and displays an imposing grandeur. The platform has three staircases and each has seven steps covered with red carpet. The central carpet bears a huge pattern of golden dragon, cloud, and water.

• 明代梳背椅

梳背椅的靠背用细圆柱均匀排列而成，如同梳子，隽秀雅致，故名。

A Comb Back Chair of the Ming Dynasty (1368-1644)

As its name suggests, the comb back chair has a comb-shaped back. Its thin, round, and curved back columns give it an elegant appearance.

凳

凳是汉代时才出现的无靠背、无扶手的坐具。目前我们见到的明式凳和清式凳大致分为长凳、方凳、长方凳和圆凳。明代凳子有方圆两类，其中以方凳的种类最多。清式凳在结构上与明式有所不同，方凳、圆凳的尺寸较明代略小些，式样清秀，宜在小巧精致的房间摆放。

Stool

The stool is a seat without back or armrest. It did not appear until the Han Dynasty (206 B.C.-220). The present-day Ming-style stools and Qing-style stools can be roughly divided into four types: long stool, square stool, rectangular stool, and round stool. Ming-style stools include square ones and round ones. Of them, the square stools have the most varieties. Qing-style stools are different from Ming-style stools in structure. The square and round stools of the Qing Dynasty are a little bit smaller than their counterparts in the Ming Dynasty. They look elegant and are suitable for small and dainty rooms.

• **清代红木长凳**

长凳是狭长的凳，又可分为条凳、二人凳和春凳三类。

A Mahogany Long Stool of the Qing Dynasty (1616-1911)

There are three kinds of long stools: bench, two-seater bench, and large bench.

• **明代黄花梨木小方凳**

方凳是凳面呈正方形的凳子，又可分为大方凳和小方凳。

A Small Yellow Rosewood Square Stool of the Ming Dynasty (1368-1644)

The square stool has a square seat and can be divided into large square stool and small square stool.

明代黄花梨木滚凳

滚凳是一种脚踏，但和一般脚踏不同，能活动筋络，促进血液循环，有利于人体健康。

A Yellow Rosewood Roller Stool of the Ming Dynasty (1368-1644)

The roller stool is a foot stool. Different from other foot stools, it can help the users to exercise their main and collateral channels, promote their blood circulation, and improve their health.

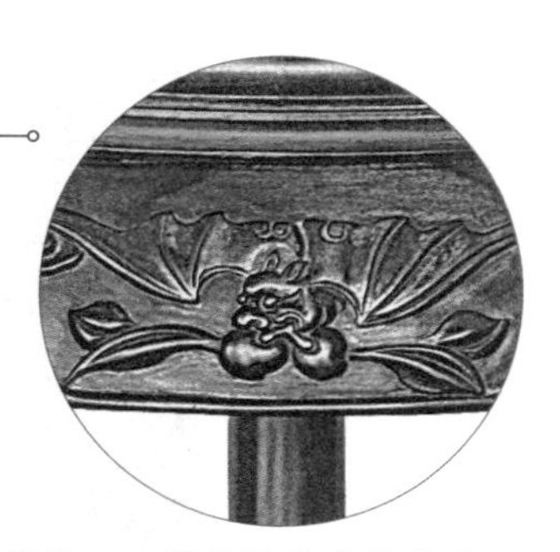

蝙蝠纹：中国传统的装饰艺术中，蝙蝠被当作幸福的象征。民间艺人运用"蝠"与"福"字的谐音赋予蝙蝠吉祥的寓意，希望幸福从天而降。

Bat pattern: In traditional Chinese decoration art, bat is a symbol for good fortune because the character of bat is a homonym of happiness. The folk artists make use of such pronunciation coincidence to invite good fortune.

清代红木圆凳

圆凳的坐面为圆形，腿足有三足、四足、五足、六足和八足之分。

A Mahogany Round Stool of the Qing Dynasty (1616-1911)

The round stool has a round seat. The round stod foot styles diverse like three legs, four legs, five legs, six legs, and eight legs.

黄花梨木

黄花梨木木质细腻，自然美观，香气持久，不易变形，坚固耐腐，是制作贵重家具和雕刻工艺品的上等材料。明清时期的黄花梨家具被视作世界家具艺术中的珍品，有温润如玉的质感、行云流水的纹理、温和内敛的色泽、淡雅芬芳的香气，非常华贵、高雅。此外，黄花梨木具有很强的韧性，在没有外力破坏的情况下，很少出现干裂现象，这也是在明式案、几中常用整块素面黄花梨木板的主要原因。

Yellow Rosewood

Yellow rosewood has a fine and smooth texture that is natural and beautiful. It sends forth lasting fragrance and does not deform easily. Being strong and performing well in resisting corrosion, it is a first-class material for making precious furniture and carving artwork. The yellow rosewood furniture in the Ming and Qing dynasties is regarded as the masterpiece among the furniture artwork in the world. With the jade-like touch, smooth grain, mild gloss, and graceful fragrance, it is truly luxurious and elegant. In addition, yellow rosewood is tenacious and seldom cracks without the impact from the outside. This is a main reason why complete pieces of raw board were used to make tables and stands in the Ming Dynasty.

- 黄花梨木切面
A Section of Yellow Rosewood

墩

墩是一种无靠背的小型坐具，坐面圆形，腹部大，上下小，造型很像中国古代的鼓，古人常在坐墩上铺锦披绣。坐墩可用草、藤、木、瓷、石等材料制成。明清时期，墩的样式较明代多，有圆形、海棠形、多角形、梅花形、瓜棱形等。从造型上看，明式坐墩造型敦实，清式坐墩比明式坐墩秀气，坐

Drum Stool

The drum stool is a small round backless seat. It has a large belly and two smaller ends, looking rather like a drum used in ancient China. The ancient people often covered it with brocade. It can be made of materials such as straw, vine, wood, porcelain, and stone. In the Ming and Qing dynasties, Ming-style drum stools were popular and they were made in many shapes such as round, plum-leaf crab, polygon, plum blossom, and melon

• 苏州园林中的圆桌和圆墩
Round Table and Drum Stools in Suzhou Gardens

明代圆形木坐墩
A Round Wooden Drum Stool of the Ming Dynasty (1368-1644)

edge. As for the design, Ming-style drum stools were stocky while Qing-style drum stools were smaller and more elegant, especially the round ones. Qing-style drum stools were also better decorated and thus suitable for larger rooms.

Folding Stool

The folding stool is a foldable square seat. Easy to carry and store, it has been widely used by ordinary people for thousands of years. The seat of the folding stool is made of ropes, velvets, or leather strips. Some folding stools even have carved seat mounted with decorative metal parts.

面也比较小，尤其是圆形坐墩。清式坐墩雕饰华美，适宜陈设在大房间，具有很好的装饰效果。

杌

杌，又称“交床”“马扎”，是一种可折叠的方形坐具，由于携带、存放方便，千百年来一直被普通百姓广泛使用。杌面多用绳索、丝绒或皮革条带等材料制成，有的杌面还施以雕刻，加金属饰件。

明代黄花梨木杌
A Yellow Rosewood Folding Stool of the Ming Dynasty (1368-1644)

> 桌案

桌和案是中国传统家具中品种最多的一类，人们常常将桌案并称。按其种类，桌案类家具可分为桌、案、几等。桌子的四条腿都在桌面的四个角上，并与桌面垂直；案的四条腿不在四角，而是往里侧缩进；几则比桌、案在形制上要小得多。

桌

桌是一种上有方形、长方形、圆形等多种形状桌面，下有腿足的家具，通常要跟椅或凳配套使用。在明清家具中，有些桌和案的形状十分相似，一般认为桌比案略小。明清时期的桌有方桌、长方桌、半桌等形式。按照用途，桌又可分为多种类型，比如放在炕上或床上使

> Table and Desk

The table and desk have the most varieties among traditional Chinese furniture. Chinese people often mention them together. This kind of furniture can be divided into table, desk, stand and so on. The four legs of a table are on four corners of the table top and are vertical to it. The four legs of a desk are not on the four corners, but recessed from the edge. The stand is much smaller than the table and desk.

Table

A table has several legs and a top in various shapes, including square, rectangle, and round. It is often used together with chair or stool. Among furniture in the Ming and Qing dynasties, some tables and desks are in similar shapes. Generally, it's been regarded

用的桌叫“炕桌”，为弹琴而设的桌叫“琴桌”，为下棋而设的桌叫“棋桌”，此外还有茶桌、酒桌、书桌、画桌等。

that table is a little bit smaller than desk. Tables in the Ming and Qing dynasties can be divided into many types, including those used on the *Kang* (*Kang* table, *Kang* is a heatable brick bed common in North China), those for drinking, those for placing a musical instrument (musical instrument table), those for playing chess (chess table), those for offering sacrifice (altar), those for holding a banquet, and those in square or rectangular shape, and half tables. There are also tables for enjoying tea, for writing (desk), and for painting.

- **清代红木方桌**

方桌是桌面为正方形的桌子，规格有大小之分：尺寸大者叫“八仙桌”，中等者叫“六仙桌”，尺寸小者称“四仙桌”。方桌是家中必备的家具，可贴墙放、靠窗放、贴着长桌案放，或居中放在室内，配置四把方椅或方墩。

A Mahogany Square Table of the Qing Dynasty (1616-1911)

The square table has a square table top. A large square table is called an eight-immortal table, a smaller one a six-immortal table, and the smallest one a four-immortal table. The square table is an indispensable piece of furniture in any household. It can be placed by the wall, by the window, by a long desk, or in the center of the room. Each table is equipped with four square chairs or square stools.

- **清代六方桌**

六方桌是桌面为正六边形、有六条桌腿的桌子。

A Six-sided Table of the Qing Dynasty (1616-1911)

The six-sided table has six legs and a table top in regular hexagon.

• 清代紫檀木长桌

长桌又称“条桌”，是桌面为长方形的桌子，桌面的长宽比超过3∶1。长桌的体积不大，可随意摆放，使用方便，是明清时期最为常用的一种桌子，深受各个阶层人士的喜爱。中国古代文人士子所用的书桌、画桌所见实物也属于长桌。

A Red Sandalwood Long Table of the Qing Dynasty (1616-1911)

Also known as the rectangular table, the long table has a rectangular table top. The ratio of its length to width,which exceeds 3:1. Not too large, a long table can be placed anywhere and is easy to use. It was the commonest table in the Ming and Qing dynasties and was loved by people at all classes. In fact, the tables used by ancient scholars and officials for study and painting were long tables.

• 清代红木书桌

书桌是用于读书、写字、作画的长方形桌子，其造型经历了从明式书案到清式书桌再到民国写字台的演化过程。

A Mahogany Desk of the Qing Dynasty (1616-1911)

The writing table (desk) has a rectangular table top. Both its design and Chinese name have been through countless evolution in history.

• **清代楸木石面月牙桌**

月牙桌是桌面为半圆形的桌子，像一弯新月，优雅怡人。两张可拼合成一个圆桌，一张可以单独靠墙放置，扩大了室内空间的利用率。

A crescent Table of the Qing Dynasty (1616-1911) Made of Chinese Catalpa Wood and with a Marble Top

The crescent table has a half-round table top, very much like a crescent. Two crescent tables can be put together to make a round table. When separated, they can be placed by the wall to make more space available.

楸木

楸木属于软木，本身并不算名贵，价格相对低廉，是最早用于做仿红木家具的木材。但随着红木的日渐稀少，楸木渐渐成为家具中的高档用材。楸木重量较轻，色泽深，材质松，棕眼大而分散，但木材的性能较稳定，不易开裂，不易变形，非常坚实耐用。同时，楸木的质地较为细腻，经打磨后，家具能呈现出亮泽和优美的木纹。楸木家具既具备红木家具的实用和美观的特点，又具备红木所达不到的稳定性能，因此受到越来越多人的喜爱。

Chinese Catalpa Wood

Chinese catalpa wood is a kind of softwood. Not precious itself, it used to be cheap and was the earliest material used to make the furniture imitating the one made of mahogany. However, with mahogany getting scarcer and scarcer, Chinese catalpa wood has gradually become a top-grade wood for making furniture. Chinese catalpa wood is light and has a dark color and a loose texture. Although it has large and scattered scars, it is rather stable and does not crack or deform easily. It is a very solid and durable wood. Meanwhile, its fine texture contributes to the shiny gloss and beautiful grain on the furniture after grinding. In fact, the furniture made of Chinese catalpa wood is not only as practical and beautiful as mahogany furniture, but more stable than mahogany furniture. It is therefore loved by more and more people.

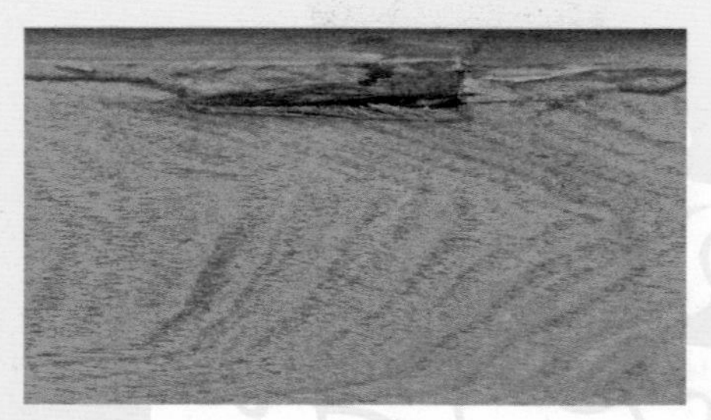

• **楸木的切面**

A Section of Chinese Catalpa Wood

- **清代棋桌**

棋桌是一种专用于弈棋的方桌或长桌。明清时期的棋桌设计巧妙，桌面能活动，一般为双层或三层。下棋时拿下一层桌面，便会露出棋盘，套面之下有暗屉，用来存放棋具。

A Chess Table of the Qing Dynasty (1616-1911)

The chess table has a square or rectangular table top. In the Ming and Qing dynasties, chess tables were of ingenious design. The table top was movable and there were two to three layers. When playing chess, a layer of table top could be taken away to expose the chess board below. Under the chess board were hidden drawers for storing chess articles.

- 桌、案、椅组合

A Combination of Desk, Table and Chairs

狮子戏球纹：狮子为百兽之王，是权力与威严的象征。狮子戏球纹是中国传统的吉祥纹样，通常以气势威猛的雄狮构成。

The pattern of a lion playing with a ball: King of the beasts, lion symbolizes power and majesty. The pattern of a lion playing with a ball is a traditional auspicious pattern in China. The lion in the pattern is usually a powerful and imposing one.

桌面：由两张半圆桌拼接而成。

Table top: It is composed of two half-round tables.

脚踏：由两个半圆拼成，透雕几何形纹理，简洁空灵。

Footrest: It is composed of two half rounds. With geometric fretwork pattern, it looks simple and neat.

- **清代红木圆桌**

圆桌是清代才开始流行的桌式，桌面为圆形，一般有六条腿。圆桌常由两张半圆桌拼成，也有独面的圆桌。清代中期以后，大圆桌非常流行，有的可围坐十几人至二十人。

A Mahogany Round Table of the Qing Dynasty (1616-1911)

The round table was not popular until the Qing Dynasty. It has a round table top and normally six legs. A round table is usually composed of two half-round tables. There are also round tables with a complete table top. After the mid Qing Dynasty, large round table became popular and some could accommodate over a dozen to twenty people.

八仙纹：八仙纹是中国古代常用的一种装饰，即由汉钟离、吕洞宾、铁拐李、曹国舅、蓝采和、张果老、韩湘子、何仙姑八位仙人形象组成的纹样。

Eight-immortal pattern: Eight-immortal pattern was a decoration often used in ancient China. It is the pattern formed by the images of eight Taoist immortals, namely Han Zhongli, Lv Dongbin, Li Tieguai, Cao Guojiu, Lan Caihe, Zhang Guolao, Han Xiangzi, and He Xiangu.

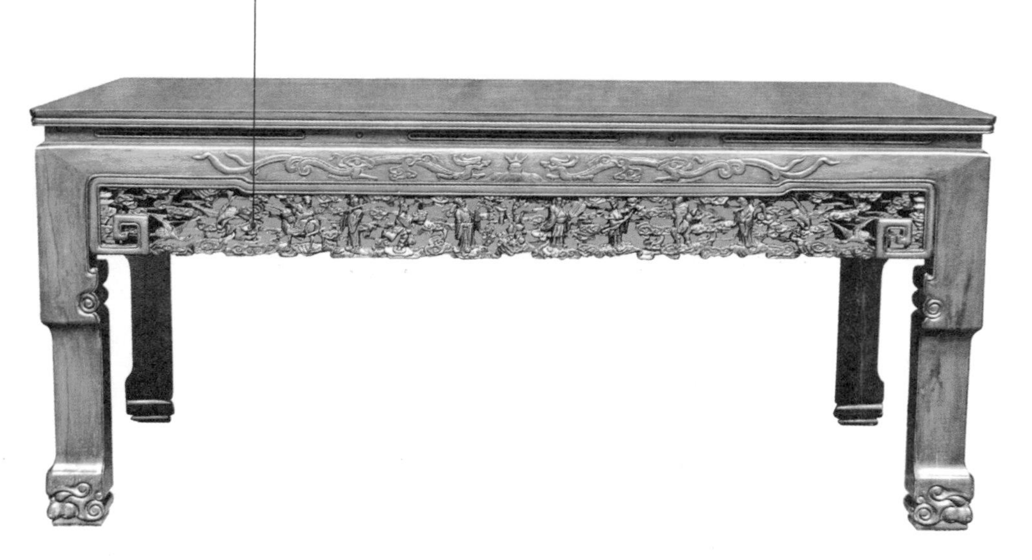

- **清代红木八仙纹供桌**

供桌是中国人在年节时供奉祖先或寺庙中用来陈设祭品，放置壶、杯、盘等祭器的桌子。

A Mahogany Altar of the Qing Dynasty (1616-1911) with Eight-immortal Pattern

The altar is used by Chinese people to offer sacrifice to their ancestors on special occasions. It is also used in temples for placing sacrifice or sacrificial utensils such as bottle, cup, and plate.

- **清代铁力木大理石面心炕桌**

炕桌是在炕上、榻上使用的矮桌，一般尺寸不大，长、宽的比例约为3：2。中国古代宫廷或官宦人家的炕桌制作十分考究，使用方式也相对固定，主要放于炕床一侧或坐榻中间。民间炕桌则讲究实用，制作比较古朴简单，使用也较灵活。

An Ironwood Marble-top *Kang* Table of the Qing Dynasty (1616-1911)

The *Kang* table is used on a *Kang* (a heatable brick bed) or couch. It is low and small, with a length to width ratio of 3:2. The *Kang* tables used in imperial palace or official households in ancient China were exquisitely made. Their uses were rather fixed. They were mainly put on one side of the *Kang* or in the center of the couch. In comparison, the *Kang* tables used by ordinary people had simpler design and were used more flexibly. Being practical was their main feature.

- **清代红木琴桌**

琴桌是抚琴时专用的桌子，形体不大，比一般桌略矮些，桌面似条桌，窄而长，桌四面饰有围板，下底由两层木板组成，其中留出透气孔，使桌子成为共鸣箱。

A Mahogany Musical Instrument Table of the Qing Dynasty (1616-1911)

The musical instrument table is used when the instrument is played. It is not big and is a little bit shorter than ordinary tables. Its table top is like that of a rectangular table, narrow and long. The table is decorated with railing boards on four sides. Its bottom is composed of two layers of board. There are breathers on the board to turn the table into a resonator.

• **明代柞木酒桌**

酒桌是在酒宴中陈置酒肴的桌子。明式酒桌是一种形制较小的长方形案，比炕桌略大些，比一般的桌子要矮一些。有的桌面下设有双层隔板，用来放置酒具。清代以后，酒桌逐渐淡出了人们的视野，因为这一时期人们喜欢多人围坐大圆桌宴饮。

An Oak Table for Drinking of the Ming Dynasty (1368-1644)

The table for drinking was used in a banquet. The Ming-style table of this kind has a rectangular table top. It is a little bit bigger than a *Kang* table and shorter than ordinary tables. Some tables have a double-layer partition board under the table top to contain drinking utensils. After the Qing Dynasty (1616-1911), people gradually forgot this kind of table because they preferred dining with more friends around a large round table.

柞木

柞木木质坚硬，剖面有明显木纹，纹理美观清晰，耐磨损，易于加工。中国古代的柞木家具制式多为美人榻、炕桌、太师椅等。柞木的成材时间漫长，一般要80年左右，因而极为珍贵。柞木家具具有清雅的风格，这与黄花梨木家具非常相似。同时，随着使用时间的增长，柞木家具外表的颜色会发生变化，由浅杏黄色变为红褐色，十分美观。

Oak Wood

Oak wood has a hard texture and is easy to process. It performs well in resisting wear. Its section shows clear and beautiful wood grains. In ancient China, oak was mainly used to make beauty's couch, *Kang* table, and master's chair. It takes a long time, about 80 years, for oak tree to grow into qualified wood. It is therefore very precious. Oak furniture has an elegant style, quite similar to that of yellow rosewood furniture. Moreover, with time passing by, oak furniture will change its color, from light apricot yellow to reddish brown. The older the furniture exists, the more beautiful it looks.

• **柞木的板材**

Oak Board

案

案是一种案面为长方形、下有足的家具。秦汉时期，案是专门用来放置食物的家具。到了魏晋南北朝时期，案又可细分为食案、书画案、奏案、香案等。两宋时期，出现了富贵家庭中厅堂里陈设的高足案。到了明清时期，案的用途变广，主要有平头案、翘头案等，大多陈设在大厅正中。

Desk

The desk has a rectangular top and several legs. In the Qin and Han dynasties, it was used exclusively to place food. By the Wei, Jin, and the Southern and Northern dynasties, desks had been used for different purposes, including those for dining (dining desk), those for writing and painting, those for handling official affairs (official desk), and those for supporting incense burners (incense altar). In the Northern and Southern Song dynasties, the tall-leg desk appeared in the hall of the rich families. By the Ming and Qing dynaties, desks had been used more widely. There were mainly flat-top desk and everted-flange desk, which were normally placed in the center of the hall.

• **清代翘头案**

翘头案的案面两端装有翘头，有的翘头与案面用一块木料做成。制作翘头案多用较厚的木料，一般镂空雕刻精美的图案。翘头案多设在厅内中堂，也可放于窗前，用来摆放花瓶或梳妆用具。

A Everted-flange Desk of the Qing Dynasty (1616-1911)

The everted-flange desk has two everted flanges on its top. Some everted flanges are made from the same piece of board as the top. Thick wood is normally used to make such desk. It often bears the exquisite fretwork patterns. Usually placed in the center of the hall or beside the window, it is used as a platform to support vases or makeup tools.

- **清代平头案**

平头案是放在书房中用来写字作画的案，案面平直，尺寸较宽大，不设抽屉。

A Flat-top Desk of the Qing Dynasty (1616-1911)

The flat-top desk is placed in the study for writing and painting. It has a wide, flat, and straight top, but has no drawer.

- **山西平遥城隍殿前的奏案**

奏案是专供帝王和官吏处理政务、接受奏章使用的，形制比一般的平头案和翘头案要大。

The Official Desk in front of the Chenghuang Hall in Pingyao Town, Shanxi Province

The official desk was used by emperors and officials to handle administrative affairs and receive reports. It is much larger than ordinary flat-top desks and everted-flange desks.

食案

食案是中国古代专用于放置食器的器具，形如方盘，多为木制，案面常见等距排列的圆涡纹，方便摆放干果和食物，四角底部有矮足。食案在先秦两汉时期比较秀气，尤其是小型食案，低矮、轻巧，案面四周设有拦水线，以防止食物汤水溢出，适合人们席地而坐时持案进食。

Dining Desk

In ancient China, dining desk was used to place dining utensils. It was like a square disk and was mainly made of wood. On its top were some round whirlpool patterns separated at an equal interval for holding dry fruits and other food. It had short legs at four corners. In the Qin and Western and Eastern Han dynasties, dining desk was small, short, and light. Stop lines were carved on the edge of the desk top to prevent overflows. It was suitable for diners sitting on the ground.

• **苏州寒山寺摆放的供案**

供案是中国古代祭祀时放置祭器的长条形案，一般木板较厚。

The Sacrifice-offering Desk (Also Called Altar) Placed in the Hanshan Temple of Suzhou

The sacrifice-offering desk has a rectangular shape. It is made of thick board and is for supporting sacrificial utensils during sacrifice offering ceremonies.

• **长沙马王堆汉墓出土的西汉食案**

此食案为长方形，木制，涂红、黑漆，样式非常精美，线条流畅，色彩明晰。漆案案底铭刻“轪侯家”三字，表明了墓主人的确切身份。

A Dining Desk of the Western Han Dynasty (202 B.C.-9 A.D.)Unearthed from the Tomb of the Han Dynesty at Mawangdui, Changsha

This dining desk is in a rectangular shape. It was made of wood and coated with red and black paints. It has an exquisite design, a smooth contour, and clear coloration. Its bottom bears three Chinese characters *Dai Hou Jia*, indicating the owner of the tomb.

几

几是一种案面狭长、下有足的矮形家具，一般设于座侧，方便人们坐时依凭和搁置物件。明清时期，几按用途分有凭几、炕几、茶几、香几、花几等。

Stand

The stand is normally low and has a narrow and long top and several legs. It is usually placed beside the seats for people to lean against or put their things on. In the Ming and Qing dynasties, stand was divided into several types, including the supporting stand, *Kang* stand, teapoy, incense stand, and flower stand.

槐木

槐木的木质较硬，木纹较匀，纹理平直，耐腐性强，心材呈深褐色或浅栗褐色，是中国北方家具常用材之一。槐木一般用来制作凳桌类家具，美观大方。此外，槐树还是中国重要的庭园和街道绿化树种，被视为吉祥树种，民间有谚语："门前一棵槐，不是招宝，就是进财。"

Locust Wood

Locust wood is hard with even and straight grains. It performs well in resisting corrosion. Its heartwood is dark brown or light chestnut brown. As one of the commonly-used wood for making furniture in north China, locust wood is normally used to make stools and desks. Locust wood furniture is pretty and graceful. Locust is also an important tree for garden and street landscaping. It has been viewed as an auspicious tree species. A proverb goes: "Planting a locust in front of your house, you end up either bringing in wealth or amassing fortunes."

• 北京北海的国槐

A Chinese Locust Tree in Beihai Park, Beijing

- **清代核桃木香几**

香几是用来摆放香炉的高腿家具，形体高大，庄重大方，样式很多。

A Walnut Wood Incense Stand of the Qing Dynasty (1616-1911)

The incense stand is for placing incense burners. With tall legs, it has a high stature and a solemn and graceful appearance. It can be made in varied styles.

- **明代红木茶几**

茶几是专门用来摆设茶具的家具，玲珑精致，与炕几有很多相似或通用之处。

A Mahogany Teapoy of the Ming Dynasty (1368-1644)

The teapoy is for placing tea set. It is dainty and exquisite and shares many similar or common features with a *Kang* stand.

- **明代槐木炕几**

炕几是一种放在床榻或炕上使用的矮形家具。从结构上看，由三块板叠加而成。

A Locust Wood *Kang* Stand of the Ming Dynasty (1368-1644)

The *Kang* stand is short. It is used on bed, couch, or *Kang*. It is composed of three crossing boards.

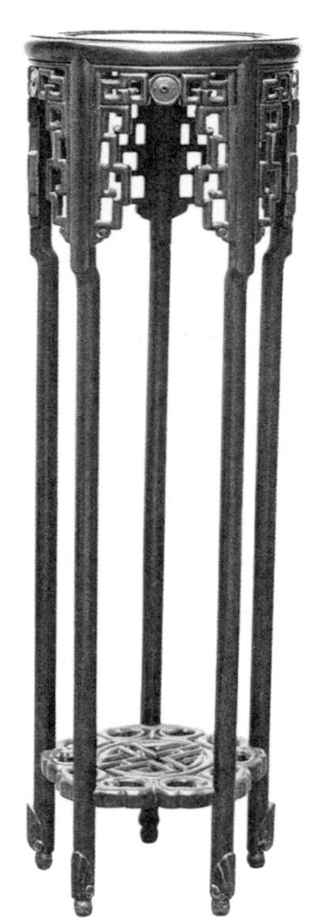

• **清代红木圆花几**

花几是摆放盆花的家具，可用于室内外，一般为方形、圆形、六角形等，有高有矮，一般成对使用。花几多选用花梨木、紫檀木等名贵的木材制作，造型高雅，设计精巧，几面常嵌大理石、玉、玛瑙、五彩瓷面等。

A Mahogany Round Flower Stand of the Qing Dynasty (1616-1911)

The flower stand is for placing potted flowers both indoors and outdoors. It normally takes the shape of a square, round, or hexagon. There are tall flower stands as well as short ones. Normally, flower stands are used in pairs. They are mainly made of precious wood such as padauk wood and red sandalwood. They have elegant shapes and ingenious design. Marble, jade, agate, and multicolored porcelain are often mounted on top of the flower stand.

核桃木

核桃木质地温润细腻，纹理美观，与花梨木十分相似，用其制作的家具富有古朴雅致的美感。此外，核桃木材质坚硬，韧性很强，非常适宜雕刻。北京的故宫博物院、山西的曹家大院和乔家大院等，现今保存有雕刻精美的核桃木家具。高档的核桃木家具一般包括床榻、几案、柜格等，用料厚实，形体宽大，做工十分精细。

Walnut Wood

Walnut wood has a mild, fine, and smooth texture, and beautiful grains, which is similar to padauk wood. The walnut wood furniture presents a simple and elegant beauty sense. In addition, walnut wood is hard and tenacious, highly suitable for carving. Many pieces of beautifully carved walnut wood furniture can be seen in the Palace Museum of Beijing, and the Cao's and Qiao's Grand Courtyard in Shanxi Province. Top-grade walnut wood furniture includes bed and couch, stand and desk, cabinet and shelf. Made of thick and solid wood, they have a wide and large build and show refined workmanship.

• **清代核桃木佛柜（局部）**

核桃木的木质纹理有些像黄花梨木。

A Buddha Cabinet of the Qing Dynasty (1616-1911) Made of Walnut Wood (Part)

The texture and grain of walnut wood is similar to those of yellow rosewood.

> 箱柜

箱柜类家具主要用于储藏、存放或搁置器物，有箱、柜、橱和架格等类型。架格和橱最早专用于存放食物，后来随着用途的逐渐扩大，出现了多种形制的柜和不同用途的橱。

柜

柜是用来存放物品的大型家具，也是居室中必备的家具。一般来说，柜的体积高大，高度大于宽度，柜门通常为对开两扇门，柜内装隔板隔层，柜门上装有铜饰件，方便开合及上锁。柜按形制分，有亮格柜、圆角柜、方角柜等类型。

> Case and Cabinet

Case and cabinet-style furniture is mainly for storing articles. It is in many types, including case, cabinet, closet, and shelf. Shelf and closet were initially used for storing food only. Later, they were put to more and more uses. Consequently, cabinets of various forms and closets of different uses came into being.

Cabinet

The cabinet is a large piece of furniture for storing articles. It is indispensable in every household. Generally speaking, a cabinet has a large volume and its height is larger than its width. It has a pair of opposite doors, both decorated with copper ornaments to facilitate the doors' opening, closing, and locking up. Inside the cabinet, there are partitioning boards to divide the interior space into cells. There are various types of cabinets, including display cabinet, round-corner cabinet, and square-corner cabinet.

亮格：柜子上没有门的开敞式隔层，可陈设古玩，放置器物，并用隔板分出隔层，后背设有栏杆，两侧用立柱固定，中间透雕花纹，后背和两侧均空敞。

Display shelf: It is the open shelf on the cabinet for displaying antiques and other articles. It can be partitioned into different layers with boards. Railing can be set in the back and is fastened with standing pillars at both sides. There are fretwork patterns in the center. Its back and two sides are all open.

抽屉：设于亮格和柜之间，上有铜拉环，可存放小件物品。

Drawer: Drawers are set between display shelf and cabinet. Equipped with copper pulling rings, they are suitable for storing small articles.

柜：立柜设在最下面，设门，有利于稳定柜子。

Cabinet: The cabinet is set at the lowest part. It has doors. The arrangement is good for stabilizing the entire cabinet.

- **现代亮格柜**

亮格柜是架格和柜子的组合，一般架格在上，柜子在下，用于陈设和收藏。架格有一层的，也有双层和多层的。亮格柜一般放在厅堂或书房，极富文人气息，深受人们喜爱。

A Modern Display Cabinet

The display cabinet is a combination of shelf and cabinet for article display and storage. Normally, the shelf is above the cabinet. Some cabinets have one layer of shelf and others have two or more layers. The display cabinet is usually kept in the hall or study. With rich cultural significance, it is deeply loved by people.

仕女图案：仕女图案是以中国古代女性为表现内容的一种传统装饰图案。明清时期的家具上广泛运用仕女图案，婀娜多姿的仕女形象是中国古代家具装饰的特色之一。

Lady figure pattern: Lady figure pattern is a traditional decorative pattern widely used on Ming Dynasty furniture and Qing Dynasty furniture. The pretty and charming lady figures form one of the characteristic decoration on ancient Chinese furniture.

柜帽：柜帽安在柜顶之上，转角呈圆弧形。

Cabinet cap: It is installed on top of the cabinet. It has round corners.

- **清式榆木红漆圆角柜**

圆角柜的顶部设有柜帽，转角多做成圆弧形，可分四扇门和两扇门圆角柜。四扇门圆角柜的形制和两扇门圆角柜大致相同，只是更宽一些。

An Elm Wood Red-painted Round-corner Cabinet in the Qing-style

The round-corner cabinet has a cabinet cap on its top. It has round corners and four doors or two doors. The four-door and two-door cabinets are in the same shape. The former is a little bit wider than the latter.

榆木

榆木的天然纹路美观，平直而粗犷豪爽，很像鸡翅木的花纹，色彩质朴，质地坚硬，易于加工，一般透雕浮雕均能运用，是古代中国人非常推崇的家具用材。榆木家具品种非常丰富，几乎包括了所有家具类型。榆木家具保留了明清家具的造型，不虚饰、不夸耀，方中带圆、自然得体、刚柔相济。

Elm Wood

The natural grain of elm wood is straight, rough, unaffected, and beautiful, very much like that of chicken-wing wood. Elm wood has an unadorned color and a hard texture and is easy to process. Suitable for making fretwork and relief, it was highly valued by the ancient Chinese as a material for making furniture. Elm wood can be used to make almost all kinds of furniture. Elm wood furniture keeps the design of Ming and Qing furniture. It has neither surplus decoration nor unnecessary design. Square with roundness, it displays a natural and appropriate form and strikes a proper balance between hardness and softness.

- **榆木的切面**

  A Section of Elm Wood

- **清代炕柜**

  炕柜是一种摆放在炕上的圆角柜，清代满族人喜欢用炕柜，因此宫廷中多见炕柜。

  A *Kang* Cabinet of the Qing Dynasty (1616-1911)

  The *Kang* cabinet is a round-corner one used on the *Kang*. Manchu people liked to use the *Kang* cabinet in the Qing Dynasty. Therefore, it was usually seen in imperial palace.

• 清代花梨木方角柜

方角柜简称“立柜”，顶部没有柜帽，上下垂直，四角见方。

A Padauk Wood Square-corner Cabinet of the Qing Dynasty (1616-1911)

Also called the standing cabinet, the square-corner cabinet does not have a cap on the top. Its entire body is vertical and its four corners are square.

• 清代核桃木箱柜

箱柜专用于放置箱子，柜顶平直整齐，下设三只抽屉，柜面的尺寸按照箱体底面的尺寸来制作。

A Walnut Wood Case Cabinet of the Qing Dynasty (1616-1911)

The case cabinet is for storing cases. It has a straight and even top and three drawers on the lower part. The cabinet size is decided based on the size of the case bottom.

顶箱：较大型的立柜顶一般都置有顶箱，形成叠式柜。

Upper cabinet: A large standing cabinet normally has a top cabinet. They form a compound cabinet together.

底柜：方正平直，外形棱角分明。

Lower cabinet: It has a square and straight shape and a clear contour.

柜门：透雕云龙纹，增强透气性，同时具有装饰性。

Cabinet door: The door has dragon pattern fretwork to guarantee better ventilation and a good decorative effect.

清代紫檀木顶箱柜

顶箱柜由顶柜和底柜两部分组成，可为一对组合排放，也可拆分为左右各一。

A Red Sandalwood Compound Cabinet of the Qing Dynasty (1616-1911)

The compound cabinet is composed of two parts: upper cabinet and lower cabinet. They can be arranged as one on top of the other or one beside the other.

橱

橱是桌案与柜的结合体，一般来说，橱比柜小些，宽度大于高度，顶部的面板既可当案来用，又可存放物品。根据形制的不同橱主要分为闷户橱和柜橱两类。

Closet

The closet is a combination of desk and cabinet. Generally, the closet is smaller than the cabinet. Its width is longer than its height. Its top board can serve as a desk or a platform for storing articles. Divided by shape, there are close-chamber closet and cabinet closet.

闷仓：设于抽屉下，仓内可以存放物品，只有拉出抽屉才能取出仓内物品，故名。

Close chamber: Set under the drawers for storing articles, it is accessible only when drawers are pulled out, hence its name.

- **明代黄花梨木闷户橱**

闷户橱是案和橱的结合，具备承置物品和储藏物品的功能。闷户橱的抽屉下设有“闷仓”，取放物品时要将抽屉拉出，有较好的隐蔽性。闷户橱是古代民间最流行的家具式样之一，多摆在内室存放日常用品。民间嫁女多用红头绳将嫁妆系扎在闷户橱上面，象征吉祥。

A Yellow Rosewood Close-chamber Closet of the Ming Dynasty (1368-1644)

The close-chamber closet is a combination of the desk and closet. It is for storing articles. Under its drawers is a closed chamber, which can be accessed only when drawers are pulled out. The close-chamber closet was one of the most popular furniture gears among the ancient ordinary people. It was often placed in inner rooms for storing articles of daily use. When a girl got married, her parents would tie her dowries to a close-chamber closet with a piece of red rope to wish her good luck.

• 明代红木柜橱

柜橱是由闷户橱演变而来的一种橱子，抽屉下没有闷仓。其抽屉以下空间设计成一个尽可能大的柜体，正面安柜门，用材多选用红木。

A Mahogany Cabinet Closet of the Ming Dynasty (1368-1644)

The cabinet closet evolved from the close-chamber closet, but it does not have a close chamber under the drawers. A big cabinet is designed under the drawers instead. There is a cabinet door on the front side. It is often made of mahogany.

• 近代佛橱

佛橱是设在家中供奉佛像的家具，一般做工都极为精美，式样较多。

A Modern Buddha Closet

The Buddha closet is for enshrining Buddha statues. It is often exquisitely made and has many styles.

架格

架格是以立木为四足，用横板将空间分割成几层，用来陈设、存放物品的高形家具。明式架格的式样和柜相似，多设有抽屉，每层亮格的后背和左右两侧多设有栏杆。清式架格较明代普及，式样和做工也都优于明式架格，是厅堂、书房之中主要的陈设。一般将左右及后面用板封闭，还多在抽屉上刻繁琐的花纹，有些花纹带有明显的西洋装饰风格。

Shelf

The shelf is tall and has four wooden legs and several layers partitioned with boards. It is for displaying and storing articles. Ming-style shelf has the similar shape as the cabinet. It usually has drawers. It also has railings on the left, right, and back sides of each display case. Qing-style shelves were more popular and showed better shapes and workmanship than Ming-style ones. They were the main furniture in the hall and study. Normally, the left, right, and back sides are covered with boards. The drawers commonly bear sophisticated patterns, which show obvious western decoration style.

抽屉：两个抽屉设在架的中部，上安铜吊牌拉手，可放置小物件。架格上设抽屉多安在便于开关处，高度一般和人的胸部相当。

Drawers: There are two drawers in the middle of the shelf and each is installed with a copper pulling ring. The drawers are for storing small articles. They are arranged at a height to the chest of a person for easy opening and closing.

亮格：共三层，上面可放置书籍、古玩、器皿等物件。

Display case: There are three layers in total for placing books, antiques, and household utensils.

• 明代红木架格

A Mahogany Shelf of the Ming Dynasty (1368-1644)

- **清代红木多宝格**

多宝格又称“博古格”，用横、竖板将架格隔成大小不同、高低错落的多层小格，专门用来陈放文玩古器，在清代非常流行。

A Mahogany Treasure Shelf of the Qing Dynasty (1616-1911)

The treasure shelf, also called *Boguge,* is composed of many small cases at different levels divided with horizontal and vertical boards. For displaying antiques, it was very popular in the Qing Dynasty.

- **苏州网师园中的书架**

书架，又称“书格”，是专用来放书籍的架格。

A Book Shelf in the Master-of-nets Garden of Suzhou

Also called book case, the book shelf is for storing books.

箱

箱是收藏物品的方形家具，多为木制，也有皮革、铁、竹制的，都配有铜饰件。明清时期的箱在用料、造型和装饰手法方面十分讲究，很有特色。中国古代的箱按用途分，有衣箱、梳妆箱、镜箱、官皮箱、书箱、扛箱、药箱、冰箱、提盒等。

Case

The case is a square-shaped piece of furniture for collecting articles. Most cases are made of wood and there are also those made of leather, iron, and bamboo. They all have copper ornaments. The cases of the Ming and Qing dynasties were special and exquisite in material using, design, and decoration techniques. The cases in ancient China included suitcase, makeup case, mirror case, travel case, bookcase, pole-carried case, drug case, cooling case, and hand-carried box.

锁匙：锁住箱子，起到保险的作用。

Lock: It is for locking up the case against unauthorized access.

- **明代黄花梨木衣箱**

衣箱是用来盛衣物的木箱，大多是长方形，上开盖，一般用樟木、柳木制作。

A Yellow Rosewood Suitcase of the Ming Dynasty (1368-1644)

The suitcase is a wooden case for holding clothes. Most of them are rectangular with a top lid. They are normally made of camphorwood and willow wood.

- **清代镜箱**

镜箱是盛放梳妆用具的匣子。

A Mirror Case of the Qing Dynasty (1616-1911)

The mirror case is for accommodating toilet set.

清代药箱

药箱是用来装药的箱子，一般体积很小，有不同的形制。有的在箱的两侧安装铜拉手，有的箱顶上有提手，便于搬动携带。箱体内设有十几个大小不同的小抽屉，可以存放不同的药品。

A Drug Case of the Qing Dynasty (1616-1911)

The drug case is for holding medicines. It is normally small and in different shapes. Some are installed with copper handles on both sides, others have another handle on the top for carrying. Inside the case are over a dozen small drawers in different sizes for storing different kinds of drugs.

明代黄花梨木提盒

提盒是带提梁的长方形箱盒，内有多层隔板，可放物品，有大、中、小三种规格。大号提盒盛的东西很多，需要两人穿杠抬行，中号提盒一人可挑两件，小的提盒可提在手中。中国古代的店铺送货上门或文人盛放文具赶考等，都用提盒。

A Yellow Rosewood Hand-carried Box of the Ming Dynasty (1368-1644)

The hand-carried box is a rectangular box with a handle. Inside it are several layers of partitioning boards for placing articles. There are large, medium, and small-sized boxes. The large-sized one can hold many things and has to be carried by two men. Two medium-sized boxes can be carried by one man with a shoulder pole. The small-sized box can be held in hand. The hand-carried box was used by shops for delivering goods or by students for holding stationery to attend examinations in ancient China.

漆家具

给木制家具涂漆是中国古典家具的一种装饰形式，历代宫廷家具就一直以漆家具为主，金銮宝座则涂以金漆。发展到明清时期，漆家具更是大展风采、品种丰富，有朱漆、描金漆、雕漆等。其中，明代宫廷作坊的雕漆最为精美，浮雕有精美的纹样，成为宫廷家具中的佼佼者。

Painted Furniture

Painting is a decorative method for Chinese classical furniture. The court furniture throughout the history had always been painted. The throne was even painted with gold lacquer. In the Ming and Qing dynasties, painted furniture reached a new height. There were rich varieties of painted furniture, including the red paint furniture, gold traced furniture, and carved lacquerware. Among them, the carved lacquerware made by court workshop in the Ming Dynasty was the most exquisite. The fretwork bears elegant patterns and makes the carved lacquerware the best of all court furniture.

- **明代官皮箱**

官皮箱是一种旅行中用来贮物的小木箱子，体积不大，制作极其精美，上方开盖。箱体上一般雕刻有喜庆吉祥图案。这个官皮箱以木为胎，表面涂漆。

A Travel Case of the Ming Dynasty (1368-1644)

The travel case is for storing articles during a trip. It is not large, but is exquisitely made with a top lid. The case body usually bears happy and auspicious patterns. This travel case is made of wood and coated with paint.

- **清代梳妆箱**

梳妆箱是存放妇女梳洗用具和化妆用品的小箱子。

A Makeup Case of the Qing Dynasty (1616-1911)

The makeup case is for storing women's toilet sets and cosmetics.

箱盖：为对开平面盖，上面透雕两个孔。

Case lid: The case has a pair of lids, on which are two fretwork holes.

提环：箱身两侧设有双提环，方便人们搬运。

Holding ring: The case has two rings on either side for easy carrying.

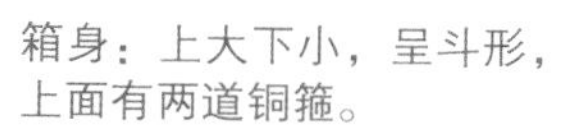

箱身：上大下小，呈斗形，上面有两道铜箍。

Case body: The upper part of the case body is larger than the lower part, showing a funnel shape. On it are two copper hoops.

底座：方凳形底座，将箱身高高地托起。

Pedestal: This is a pedestal in square stool shape. The case body rests high on it.

- **清代冰箱**

中国古代就已经有冰箱了，它是一种保存食物的木箱，里面是锡，放有天然冰块（古代冬天会把冰块储存在冰窖里，供夏天使用），外部设有铜箍。

A Cooling Case of the Qing Dynasty (1616-1911)

The ancient Chinese had invented the cooling case for storing food. It had a tin lining inside and a copper hoop outside. Natural ice cakes collected in winter were taken out of the cellar and put into the case for cooling in summer.

> 其他木家具

> Other Wooden Furniture

屏风

屏风是一种古老的家具，有分隔室内空间、挡风、屏蔽视线、装饰等诸多用途。屏风分为落地屏风和带座屏风两大类。落地屏风是多扇折叠屏风，多为双数，最少两扇，多可达数十扇。带座屏风是把屏风插在底座上，多为单数，三、五、七、九各数不等。每扇屏风下边框的两侧都有腿，插入底座的孔中，屏顶大多有装饰，或雕刻，或镶嵌，或绘画，或书法。明代后期出现一种挂在墙壁上的挂屏，也用于装饰。清代屏风品种繁多，出现了“炕屏”“寿屏”等屏风，形体雄大，屏心常镶大理石、玻璃等饰品。

Screen

The screen is an ancient piece of furniture with many uses, including room space partitioning, wind shielding, sight screening, and decoration. There are two major types of screens: floor screen and pedestaled screen. The floor screen is a multi-panel screen and its panels are normally in even numbers. There are at least two panels. Some screens can have as many as several dozen panels. The pedestaled screen is inserted in a pedestal and its panels are normally in odd numbers, such as three, five, seven, and nine. Each screen has legs on both sides of its lower frame, which will be inserted into the holes on the pedestal. Screen tops are usually decorated with carving, inlaying, painting, or calligraphy. In late Ming Dynasty (1368-1644), a wall screen emerged, which was for decoration. In the Qing Dynasty (1616-1911), there were numerous screen varieties. New types of screens came into being, like the *Kang* screen and birthday scroll screen, which was massive with ornaments such as marble and glass on the central panel.

清代镂雕福寿纹砚屏

砚屏是放在书桌、画案上供欣赏的小型工艺屏风，形制小巧，做工精美。此砚屏雕刻的福寿纹是一种吉祥纹样，由蝙蝠、寿桃或团寿组成，借助“蝠”与“福”的谐音表现福寿吉意。

An Inkstone Screen with Carved Painting of Good Luck and Longevity of the Qing Dynasty (1616-1911)

The inkstone screen is a small-sized artistic screen placed on desk for decoration and appreciation. It displays exquisite workmanship. The pattern of good luck and longevity is composed of the images of bat, birthday peach, or the round Chinese character of longevity to indicate the auspicious meaning.

清代红木嵌瓷山水挂屏

挂屏就是一种悬挂在墙壁上的屏风，属于工艺装饰画，挂在室内墙上代替卷轴画。清代挂屏的装饰技法丰富多彩，在宫廷后妃居住的寝宫里处处可见。挂屏一般成对或成组使用，也有单屏。

A Mahogany Wall Screen Inlayed with a Porcelain Landscape Painting of the Qing Dynasty (1616-1911)

The wall screen is hung on the wall to replace scroll painting. An artistic painting of rich decorative techniques, the wall screens of the Qing Dynasty were a common decoration in the imperial resting palaces. They are often used in pairs or groups while there is also solo screen.

透雕：屏心和屏框之间透雕出空灵精美的几何图案。

Fretwork: There are fancy geometric fretwork patterns between the central panel and screen frame.

屏心：镶嵌天然大理石，自然的石纹形成一幅极具水墨意境的山水画。

Central panel: It has an inlay of natural marble. Its natural stone grain forms a landscape painting of Chinese ink and wash.

上屏板：上屏板浮雕博古纹，即瓷、铜、玉、石等各种古器物的纹理，也有添加花卉、果品作为点缀的，寓意清雅高洁。

Upper screen board: The upper screen board bears old relief grains, namely those on antique porcelain, copper, jade, and marble. There are also patterns of flowers and fruits to symbolize elegance and loftiness.

下屏板：下屏板浮雕“福禄寿”纹。福、禄、寿在中国民间被认为是天上三吉星，“福”寓意五福临门，“禄”寓意高官厚禄，“寿”寓意长命百岁，因此“福禄寿”纹象征幸福、吉利和长寿。

Lower screen board: The lower screen board bears relief patterns of the gods of luck, wealth, and longevity in Chinese folklores.

底座：雕蝙蝠、“寿”字等纹饰。

Pedestal: It is carved with the patterns of bat and the Chinese character for longevity.

- **清代鸡翅木座屏**

座屏又叫“插屏”，即插在屏座之上，做工精美，为陈设欣赏品，多陈设在室内主要座位的后面，用以体现主人的尊贵；也有的座屏摆放在室内进门处，起遮挡的作用。此屏风端庄典雅，呈现出鸡翅木栗褐色的色泽、绚丽的纹理。

A Chicken-wing Wood Pedestaled Screen of the Qing Dynasty (1616-1911)

The pedestaled screen is also called the inserted screen because it is inserted into its pedestal. Exquisitely made, it is an ornament for indoor decoration. Most screens of this type are placed behind the main seat to represent the owner's high status. Some are also placed at room entrance to play a sight blocking role. This screen is stately and graceful, perfectly displaying its chestnut brown color and gorgeous grain of chicken-wing wood.

屏心：由十二扇屏面组成，屏面由纸绢做成，上绘人物山水画。

Central panel: It is composed of twelve screen boards made of paper silk bearing figure and landscape paintings.

屏框：多用较轻质的木材做成，便于摆放、折叠和收藏。

Screen frame: It is mainly made of light wood for easy application, withdrawal, and storage.

- **清代黄花梨木曲屏**

曲屏又名“围屏”“折屏”，是一种可以折叠的多扇屏风，落地摆放，有两扇、四扇、六扇、八扇、十二扇等样式。曲屏为临时性陈设，摆放位置随意。摆放时，扇与扇之间形成一定的角度便可摆立在地上。为了营造某种氛围，或体现地位的高低，经常用曲屏来重新划分室内的空间，以增强每个空间的相对独立性。曲屏还可以围在床榻旁，用以挡风。

A Yellow Rosewood Curved Screen of the Qing Dynasty (1616-1911)

Also called the folding screen, the curved screen is a foldable multi-panel screen placed on the floor. There are two-panel, four-panel, six-panel, eight-panel, and twelve-panel styles. As an interim furnishing, the curved screen can be placed anywhere in a room. So long as its panels form certain angles, the screen can stand up. To create a certain atmosphere or to represent status, the curved screen is often used to repartition the indoor space and enhance the relative independence of each space. It can also be put around a bed or a couch to shield the wind.

鸡翅木

鸡翅木以美丽独特的纹理著称，它的纹理好像鸡的翅纹，交错、清晰、颜色对比强烈，且手感十分平滑。明代鸡翅木家具做工比较复杂、讲究，十分吝惜材料，一般的家具都做得较小。传世的鸡翅木家具有鸡翅木画案、翘头案、圈椅、小箱等。清代时，鸡翅木家具开始注重雕工，木纹和雕工并重。

Chicken-wing Wood

Chicken-wing wood is known for its unique and beautiful grains, like that of chicken wings. In addition, it looks colorful and touches smooth. The chicken-wing wood furniture of the Ming Dynasty (1368-1644) shows sophisticated and ingenious workmanship. The carpenters at that time begrudged the use of wood and made the furniture in smaller sizes. The chicken-wing wood furniture that has passed down includes painting desk, everted-flange desk, round-backed armchair, small case, and so on. In the Qing Dynasty, carving became another focus on the chicken-wing wood furniture besides wood grain.

• 鸡翅木的心材
Heartwood of Chicken-wing Wood

架具

架具是一种立体的用来支撑承物的家具，包括灯架、衣架、盆架、帽架、镜架、鸟架、笔架、鱼缸架、兵器架和乐器架等品种，其中衣架、盆架、灯架和镜架是中国传统家具中较为典型的架具品种。

Stand Tools

The stand tools are for supporting articles. They include lamp stand, clothes stand, basin stand, hat stand, mirror stand, bird stand, pen stand, fish bowl stand, weapon stand, and musical instrument stand. Among these, clothes stand, basin stand, lamp stand, and mirror stand are the most typical stand tools among traditional Chinese furniture.

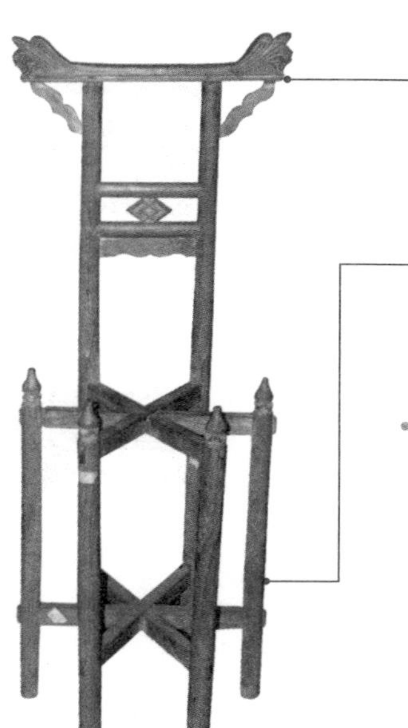

巾架：与衣架形制相似，多为木制，与盆架组合使用，用于悬挂毛巾。

Towel stand: It has a shape similar to that of a clothes stand. Normally made of wood, it is used together with a basin stand for hanging towel.

六腿足：“米”字形结构，有两层，上层放盆。

Six-leg stand: It has a crisscross structure and two layers. The upper layer is for placing the basin.

- **明代高盆架**

盆架是用于放置铜洗脸盆的家具，分高、矮两种。高盆架是巾架和盆架的结合，多为六足，最里面的两足加高成为巾架。矮盆架腿足等高，不设巾架，有些可以折叠。

A Tall Basin Stand of the Ming Dynasty (1368-1644)

The basin stand is for placing copper basins. A tall basin stand with six legs is the combination of the towel stand and basin stand. The innermost two legs can be lengthened to form a towel stand. The legs of a short basin stand are equally tall. Without the towel stand, some of the short stands are foldable.

托盘：设于灯杆顶端，用来放蜡烛或油灯，上面设有灯罩。

Supporting tray: It is set on top of the lamp pole for placing candle or oil lamp. It is equipped with a lamp cover.

灯杆：方形，在底座上钻一孔，将方形灯杆插入孔中。

Lamp pole: It is a square pole inserted in the hole drilled on the bottom.

底座：典型的“十”字形底座。

Pedestal: It is a typical cross pedestal.

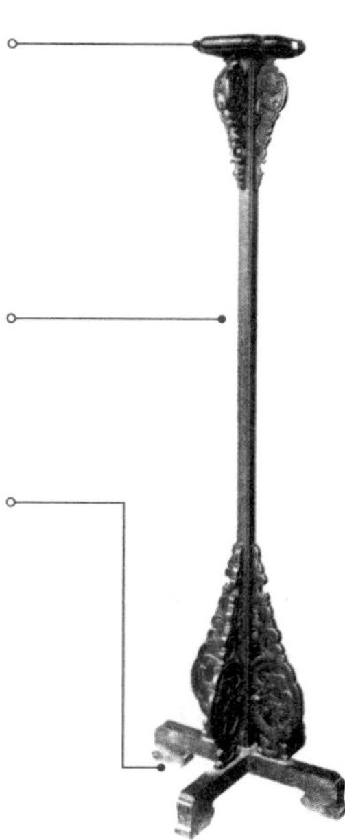

- **明代灯架**

灯架是中国古代专用来承放油灯或蜡烛的家具，多为木制，摆放随意，有装饰作用。

A Lamp Stand of the Ming Dynasty (1368-1644)

The lamp stand was for supporting oil lamp or candle in ancient China. It is often made of wood and can be placed anywhere. It has a certain decorative value.

鲁班锁

鲁班锁是中国古代一种奇巧的木玩具。相传古时候，鲁班为了测试儿子是否有资格继承自己的绝技，用6根木棍制作了一件可拼可拆的玩具，让儿子拆开再拼起来。儿子摆弄了一夜，终于成功了。这种玩具于是被后人称为“鲁班锁”，也称“孔明锁”。鲁班锁种类很多，表面上看似简单，却是一种三维的高难度益智玩具，内部的凹凸部分设计十分巧妙，凝聚着中国人不平凡的智慧。

Lu Ban Lock

The Lu Ban Lock is a magical wooden toy invented in ancient China. According to the legend, Lu Ban, the greatest master of carpentry in China, wanted to find out if his son was eligible for inheriting his skills. He made a detachable toy with six wooden sticks and ordered his son to detach and reassemble it. His son succeeded after a-whole-night trying. The toy was later called Lu Ban Lock or Kong Ming Lock. (Kong Ming was another smart figure in Chinese legend.) There are many kinds of Lu Ban Lock. It looks simple, but is actually a difficult 3D toy for improving intelligence. Its internal contour is ingeniously designed, fully demonstrating the extraordinary wisdom of Chinese people.

• 鲁班锁的一种
A Kind of Lu Ban Lock

• **明代红木衣架**

衣架是用于搭衣服的木架，有支架和横杆，是卧室中的附属家具，通常放在卧室床榻旁边或进门的一侧，并与床、桌、椅等室内家具在风格和尺寸上协调一致。

A Mahogany Clothes Stand of the Ming Dynasty (1368-1644)

The clothes stand is a wooden frame for placing clothes. Composed of a rack and a horizontal bar, it is an auxiliary piece of furniture in a bedroom. It is often placed by the bed or beside the door. Its style and size should fit in well with that of the indoor furniture such as the bed, table, and chair.

• **清代金漆镜架**

镜架是中国古代用来承托镜子的架子，一般都很小巧，结构也比较简单。

A Gold Lacquer Mirror Stand of the Qing Dynasty (1616-1911)

The mirror stand was used to support mirror in ancient China. It is normally small and exquisite with a simple structure.

玲珑剔透的木雕
Dainty and Exquisite Wood Carving

木雕是中国一种非常有特色的艺术品，由工匠选取合适的木材雕刻而成。木雕的魅力在于图案设计严谨，雕刻巧夺天工，选料考究和谐，尤其是丰富多彩的雕刻图案，赋予木雕以强大的生命力，使之更具有艺术性和观赏性。木雕图案的题材非常广泛，有人物、动物、植物、器物及山、水、云，以及传说中的瑞兽，如龙、凤、麒麟等等。

木雕多选用质地细密坚韧、不易变形的木材，如楠木、紫檀木、樟木、柏木、银杏、红木、黄杨木等。中国的木雕制作历史悠久，早在商周时期便出现了雕琢成器的木制品。到了明清时期，木雕艺术更是大放异彩，名家辈出，精品频现。按照用途，木雕可分为摆件及文房用具、建筑雕刻、家具雕刻等。

The wood carving is a characteristic artwork of China. Different works are made with different kinds of wood. The beauty of the wood carving lies in its rigorous pattern design, skillful carving workmanship, and choosing of appropriate materials. The diversified carving patterns, in particular, give wood carving great vitality and make it more valuable in both art and appreciation. Patterns of wood carving vary widely from human, animal, and plant images to mountain, water, and cloud shapes, and even include the images of some legendary auspicious animals such as dragon, phoenix, and kylin.

Wood carving is often done with the wood that has a compact and tenacious texture and does not deform easily, such as Phoebe zhennan, red sandalwood, camphorwood, cypress wood, gingko wood, mahogany, and boxwood. Wood carving has a long history in China. The earliest carved woodwork appeared in the Shang and Zhou dynasties. By the Ming and Qing dynasties, the art of wood carving had reached a new height with masters coming forth in large numbers and masterpieces being created at a high frequency. By their usage, works of wood carving can be divided into display furniture, stationery, architectural carving, and furniture carving.

> 器木雕

佛教造像和道教造像

中国历史上，佛教文化和道教文化对工艺品影响深远。尤其到了明清时期，佛道造像空前发展，类型丰富多彩，人物形象生动逼真，表情多变，衣纹清晰，制造手法技巧精湛。在皇家和民间寺庙中，皆有佛教造像木雕；另外，在民间各道观、民居、园林中也随处可见八仙、暗八仙、福禄寿等道教吉祥纹样的木雕。

> Utensil Wood Carving

Buddhist Statue and Taoist Statue

In Chinese history, Buddhist culture and Taoist culture have had far-reaching impacts on artwork. In the Ming and Qing dynasties, Buddhist and Taoist statues had an unprecedented development and they were in diversified types, looking vivid with varied expressions. Their clothes also look real, indicating the extraordinary carving skills and techniques. For royal and folk temples, the wood carving Buddha statues are indispensable. In addition, the wood carvings with some auspicious Taoist patterns such as eight immortals, hidden eight immortals, good luck, wealth, and longevity can be seen in Taoist temples, folk houses, and gardens.

- 清代寿星木雕

A Wood Carved Statue of the God of Longevity of the Qing Dynasty (1616-1911)

- 宋代漆金彩绘菩萨像木雕

A Gold Lacquer Wood Carved Bodhisattva statue of the Song Dynasty (960-1279)

- 北京雍和宫内的佛像木雕

北京雍和宫主殿万福阁是雍和宫最高大的建筑，高23米，正中一尊高大的弥勒佛立像由一整棵直径3米的白檀木雕成，通体饰金，十分壮观。

The Wood Carving Buddha Statue in the Palace of Harmony, Beijing

The Hall of Many Fortunes is the main hall and the tallest building in the Palace of Harmony, Beijing, which is 23 meters high.It houses a giant standing statue of Maitreya carved with a complete piece of Asiatic sandalwood some three meters in diameter. Completely covered with gold lacquer, the statue is truly magnificent.

• 金代漆金彩绘菩萨像木雕
A Gold Lacquer Wood Carved Bodhisattva statue of the Jin Dynasty (1115-1234)

• 财神樟木雕
A Camphorwood Caved Statue of the God of Wealth

樟木

樟木纹理细致，木质坚硬，干燥后不易变形，易于雕刻，具有强烈的樟脑香气，一般用来制作高档家具、木箱、雕刻及工艺品，也被大量应用于建筑及造船。由于樟木天然具有防蛀、防霉和杀菌等特点，因而被古代中国人制成衣橱箱柜，来贮藏棉、毛服饰等物品，使棉毛织物能免受霉菌和虫蚁的侵害。此外，古典家具的雕花板一般都是用樟木制作的。

Camphorwood

Camphorwood has fine grains and a hard texture and does not deform easily after drying. It is suitable for carving. With an intense camphor fragrance, it is often used to make top-grade furniture, case, carving, artwork, building, and boat. As camphorwood is a mothproof, mold-resisting, and germicidal wood, it was made into wardrobes, cases, and cabinets by the ancient Chinese for storing their cotton and fur clothes against mold and insects. In addition, the carved boards of classical furniture were all made of camphorwood.

• 樟木树干表皮
Bark on Camphor Tree Trunk

摆件及文房用具

木雕器件一般是小型文房用具和案头工艺摆件，如笔筒、臂搁、花插、印匣、砚盒、小瓶、小盒、杯、盅以及雕像等。这类木雕以明清时期的小件居多，且工艺高超，具有较高的审美价值和优雅的文人气息。

Display Furniture and Stationery

Wood carved utensils normally include small stationery and display furniture placed on desks such as pen-holder, armrest, pin-holder, seal container, ink stone box, vase, small box, cup, and statue. These kinds of wood carving were popular in the Ming and Qing dynasties. They all have displayed excellent workmanship, high aesthetic value, and a graceful taste of culture.

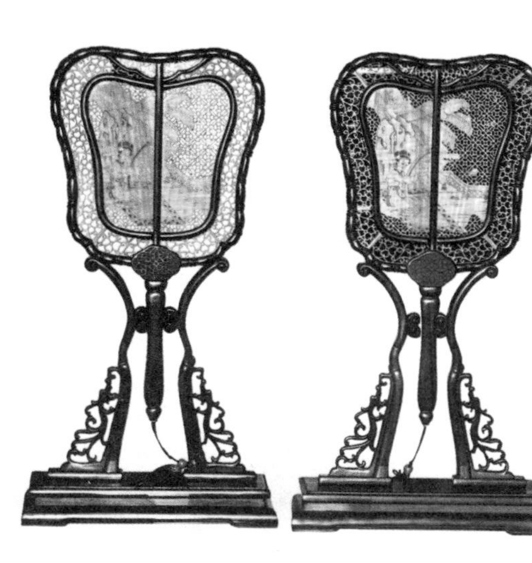

- **清代檀香扇**

扇骨采用檀香木制成。这种扇子多为妇女所用，所以特别小巧玲珑，华美精致，加上檀香木气味芬芳，盛夏酷暑颇有清心醒脑之效，兼具防止虫蛀衣服的功用。

A Sandalwood Carved Fan of the Qing Dynasty (1616-1911)

The fan skeleton was made of sandalwood. Mainly for women, this kind of fan was small and exquisite. With the fragrance of sandalwood, it can clear away the heart fire and restore consciousness in hot summer. It can also protect the clothes from worm damage.

- **黄杨木雕笔筒**

此笔筒为黄杨木制，将古梅树雕刻得叶肥枝壮，梅花怒放，形态自然娇媚，是一件木雕精品。

A Boxwood Carving Pen-holder

This pen-holder was made of boxwood. On it is a carved plum tree with thick leaves, strong branches, which also has natural and graceful plum blossoms. It is indeed a wood carving masterpiece.

黄杨木

黄杨木木质光洁，纹理细腻，异常坚韧，非常适宜精雕细琢。黄杨木雕发源于浙江乐清，成品颜色悦目、均匀，具有象牙般的色泽。此外，随着时间的推移，黄杨木雕色泽会逐年加深，给人以古朴、文雅之感。黄杨木雕的内容题材大多为中国民间神话传说中的人物，比如八仙、寿星、关公、弥勒佛等，但传世极少。

Boxwood

Boxwood has a glossy texture and fine and smooth grains. It is extremely tenacious and is therefore suitable for precision carving. Boxwood carving originated from Yueqing of Zhejiang province. It has an even and pleasant color and an ivory-like tinct. In addition, with time passing by, the color of boxwood carving will turn darker and guarantee a feeling of simplicity and elegance. Boxwood carving often depicts the figures from Chinese folk myths and legends, including the eight immortals, God of Longevity, Master Guan and Maitreya. However, few pieces of such artwork have been passed down.

- **紫檀木雕祥云纹枕**

云纹是明清时期常用的一种纹样。古代社会以农耕为主，农业生产全靠雨露滋润，无云便无雨，无雨则谷不生，因而古人由求雨转而敬云。雕刻中所用的云纹有敬天、高升、如意等吉祥寓意。

A Pillow With Cloud Pattern Carved with Red Sandalwood

The cloud pattern was frequently used on artwork in the Ming and Qing dynasties. The ancient farming society relied heavily on timely rains. Without cloud, there would be no rain. Without rain, there would be no satisfactory crop growth. Therefore, the ancient people craved for rain and then respected the cloud. The cloud pattern represents their respect for the heaven, their wish for promotion, and their will for fulfillment.

- **清代乾隆黄杨木人物摆件**

A Boxwood Human Figure Display Piece Made in the Qing Dynasty during the Reign of Qianlong (1736-1795)

清代紫檀木梅花纹葫芦瓶

A Red Sandalwood Carved Calabash Bottle with Plum Grains of the Qing Dynasty (1616-1911)

檀香木

檀香木的香味独特浓烈，有驱虫的功效，多用作书画卷轴、香料、香扇等。檀香木还用于佛教造像、佛具和佛香。中国古代，檀香木一般用于隆重的佛教及皇家庆典中，它被制作成木雕艺术品，代表专注的意念和至诚至圣的愿望，在很大程度上是一种权利与地位的象征。同时，檀香木香气醇厚，经久不散，因此檀香木工艺品更可谓贵重无比。

Sandalwood

Sandalwood has a unique and intense fragrance that can help dispel worms. It can be used to carve painting scroll and artwork and to make spice and fragrant fan. Sandalwood is also used to build Buddha statue and make Buddhist articles and incense. In ancient China, sandalwood was normally used during grand Buddhist and royal ceremonies and for making wood carving artwork. It represents the concentrated thought and the most sincere wishes. It is a symbol of power and status to a great extent. Meanwhile, its strong and lasting fragrance make the artwork carved with it even more precious.

檀香木原木

Sandalwood Logs

木俑

中国古代的丧葬风俗是土葬。死者的亲人们会为其准备一些殉葬品，一同埋在地下。其中有一种殉葬品叫“俑”，各种材料都有，但以木质和陶质为最多。木俑就是用木材雕刻成人形，使其陪伴在死者身旁。木俑按照不同的人物形态，可以分为站立俑、坐俑、骑马俑、下跪俑、划船俑、赶车俑等。木俑表现的题材很广泛，有侍女、士兵、厨师、农夫，还有禽畜俑和表演节目的乐舞俑等。木俑是雕刻非常精美的装饰品，人物形象生动，艺术水平很高。

Wooden Figurine

In ancient China, burial in the ground was the traditional practice. Meanwhile, the relatives of the deceased prepared some sacrificial objects to be entombed with him or her. One of the sacrificial objects was called figurine, which was made of all kinds of materials, but mostly of wood and earth. The wooden figurine was buried beside the deceased and it could be in many gestures, including standing, sitting, riding, kneeling, rowing, and cart driving. The human-shaped figurines could represent various professions, including maid, soldier, cook, and farmer. There were also bird and beast figurines and dancer figurines. Wooden figurines are exquisite decorative articles. They are vividly carved and have a high artistic value.

- **湖南长沙马王堆汉墓出土的彩绘木俑**

  这几件木桶以歌乐侍仆为原型，均雕刻出人的肢体、衣纹的细部轮廓，然后施以彩绘，造型简洁、比例恰当、形神兼备，可见汉朝时期木雕工艺就已经具有很高的水平了。

  Painted Wooden Figurines Unearthed from the Tomb of the Han Dynasty at Mawangdui, Changsha, Hunan Province

  These wooden figurines are those of the dancer, singer, and servant. Their bodies and clothes are carved in detail and painted vividly. With clean design, appropriate proportion, and vivid features in both body and spirit, they bear witness to the great wood carving workmanship in the Han Dynasty (206 BC-220).

木偶与面具

木偶与面具在中国历史上被广泛应用于狩猎、祭祀、战争、丧葬、舞蹈、戏剧等方面，与古代的宗教信仰密不可分。木雕面具制作工艺很复杂，一般选用纹理细腻、质地轻柔、不易变形的白杨木、丁香木等材料制成。木雕面具造型丰

Puppet and Mask

The puppet and mask have been widely used in hunting, sacrifice offering, war, burial, dancing, and play throughout Chinese history. They were closely connected with religious belief in ancient China. Wood carving mask involves sophisticated workmanship and is normally made of white poplar wood and

- 貂蝉像面具

貂蝉是三国时期的一个重要人物，也是中国古代四大美女之一。

A Diao Chan Portrait Mask

Diao Chan was an important figure during the Three Kingdoms Period (220-280). She was also one of the top four beauties in ancient China.

- 糜竺像面具

糜竺是三国时期蜀汉的一个重要官吏。

A Mi Zhu Portrait Mask

Mi Zhu was an important official in the Kingdom of Shu Han during the Three Kingdoms Period (220-280).

富，有些以龙凤作装饰，男用龙，女用凤；有些以名人作装饰，如岳飞、薛仁贵等。一般男性角色刀法明快，有棱有角，轮廓分明，而女性角色则力求娇艳妩媚，娴静高雅。

木偶是中国古代丧葬活动中由巫师来操纵的道具，以福建泉州木雕最为著名。泉州木雕具有浓厚的地方色彩和艺术风格，造型有佛像艺术雍容含蓄的风格。

clove wood, which have fine and smooth grains and a soft texture, and do not deform easily. Wood carving masks are of rich designs and some are decorated with dragon and phoenix patterns. Dragon pattern is for men's mask and phoenix pattern for women's mask. Some others are decorated with the patterns of famous figures in history, like Yue Fei and Xue Rengui. The male patterns demonstrate straightforward carving techniques, clear edges and corners, and clean contour. The female patterns display charm, elegance, and grace.

Chinese puppets were props controlled by wizards during ancient burial activities. The most famous are those made in Quanzhou, Fujian Province. Quanzhou puppet carving show rich local features and artistic style. Its design involves the artistic touch with Buddha statue.

• **木偶“白蛇”和“青蛇”**

“白蛇”和“青蛇”是中国古代神话故事中的两个人物，经常在木偶戏中出现。

White Snake and Green Snake Puppets

White Snake and Green Snake are two figures in ancient Chinese myths. They often show up in puppet plays.

> 建筑雕刻

中国的建筑木雕有数千年历史，早在秦汉时期，木雕装饰艺术的运用就已非常普遍，发展到明清时期，木雕装饰达到了巧夺天工的境界。现在所能见到的中国古代建筑木雕主要是明清两代的遗存。这一时期是中国古典建筑艺术发展的巅峰，木雕的题材丰富，工艺精湛，除了宫殿建筑以外，还有许多民居、祠堂、寺庙等，也是能工巧匠施展木雕技艺的地方。

用于古建筑装饰的木雕，主要是一些木雕构件，如梁架、屋檐、柱板、门雕和窗棂、窗饰等。雕刻手法多样，有浮雕、镂雕、透雕等，图案主要分为奇珍异兽、松竹梅兰、吉祥图纹以及各类神仙和戏曲传说人物等。在中国古典建筑

> Architectural Carving

Chinese architectural carving has a history of several thousand years. As far back as in the Qin and Han dynasties, wood carving decoration were widely used. By the Ming and Qing dynasties, it had been so excellent that it could even rival the creation of the nature. The ancient architectural carvings we see today are mainly the relics from the Ming and Qing dynasties. In fact, Chinese classical architectural art reached the peak in the Ming and Qing dynasties. Wood carving at that time involved rich topics and superb techniques. Apart from palatial buildings, the countless folk houses, ancestral halls, and temples were all the places for the skilful artisans to show their wood carving feat.

The wood carvings for ancient building decoration were mainly the

中，木雕被融入木构件，精雕细琢，使建筑和装饰完美地结合在一起，形成了一种独具特色的艺术形式。

components such as beam mount, eaves, post, door carving, window lattice, and window decoration. There are diversified carving techniques, such as relief, openwork, and fretwork. The patterns used mainly include that of rare and precious animals, pine, bamboo, plum, and orchid, various deities, and figures in operas and legends, and auspicious patterns. In classical Chinese building, wood carving is merged into many wooden components. Finely carved, they perfectly combine the building with the decoration and give rise to a characteristic art form.

• 大门上的金漆木雕

金漆木雕一般用樟木或杉木进行雕刻，然后涂上生漆，再在木材表面贴以金箔。金漆木雕以潮州木雕最为著名，多以梅兰竹菊、神禽异兽、鱼虾虫蟹、神仙传说以及潮剧故事等为题材。

Gold Lacquer Wood Carving on a Gate

Gold lacquer wood carving is normally made with camphorwood or fir wood, painted with raw lacquer, and then coated with gold foil. The one made in Chaozhou is the most famous. The carving patterns mainly include plum, orchid, bamboo, chrysanthemum, divine birds and strange animals, fish, shrimp, insect, crab, figures in myths and legends, and stories in Chao-Play.

皇家建筑木雕

皇家建筑包括宫殿、园林、寺庙等，代表着中国古代建筑艺术的最高典范，木雕艺术自然也不例外。几乎每一处皇家建筑，都会有精美的木雕装饰，使这些建筑更加富丽堂皇、气势恢弘。皇家建筑木雕的内容大多是龙凤纹或其他祥瑞鸟兽以及百花仙草等。以北京故宫为例，每一个角落都能看到精美绝伦的木雕，使整个建筑群显得高贵庄重、古朴典雅。

Royal Architectural Wood Carving

Royal buildings include palaces, gardens, and temples. They represent the highest standard of ancient Chinese architectural art. The art of wood carving is no exception. Almost all royal buildings have the exquisite wood carving decorations, which make the buildings more majestic, luxurious, and imposing. Royal architectural wood carving depicts the patterns of dragon, phoenix, auspicious birds and animals, flowers, and divine grasses. Take the Imperial Palace in Beijing (the Forbidden City) for example, beautiful wood carvings can be seen at every corner and they make the entire palace noble, majestic, and glorious.

北京故宫皇极殿外的龙纹浮雕

北京故宫的许多浮雕采用龙纹，因为龙纹是帝王专用的纹饰。明代雕刻的龙纹，龙头上的毛发大多从龙角一侧向上高耸，呈怒发冲冠状；清朝康熙年间的龙纹，龙头呈披头散发状；乾隆年间的龙纹，龙头顶上有七个圆包，正中的一个稍大，周围的略小；清朝后期，龙身臃肿呆板，毫无生机。

The dragon pattern relief outside the Hall of Imperial Supremacy of the Imperial Palace, Beijing

Many reliefs in the Imperial Palace of Beijing (the Forbidden City) have the dragon pattern, the exclusive decorative pattern for emperor. In the dragon pattern made in the Ming Dynasty, the hair on dragon head stands high along the dragon horn. The dragon pattern made during the reign of Kangxi in the Qing Dynasty has a dragon head with loose hair. In the dragon pattern made during the reign of Qianlong, the dragon head has seven round lumps, the central one a little bit larger and the others smaller. In late Qing Dynasty, the dragon became fat, rigid, and dull.

- **北京故宫养性殿内的木制蟠龙藻井**

藻井是一种高级的室内木雕装修，故宫的建筑中只有皇帝御朝理政的宫殿内才装藻井。“藻”即水藻，代表水；“井”为天文上所称的“东井”，是用来贮水的。

The Wooden Dragon Caisson Ceiling in the Hall of Spiritual Cultivation of the Imperial Palace, Beijing

The caisson ceiling is a first-class indoor wood carving decoration. In the Imperial Palace (the Forbidden City), only the halls used by the emperor to handle state affairs are decorated with caisson ceiling. The algae pattern represents water. The well shape represents the eastern well in astronomy, which is for storing water.

上部的祥云腾龙纹：云纹层层叠加，腾龙时隐时现。

Cloud and dragon grain at the upper part: The cloud pattern piles up layer by layer. The flying dragon appears here and disappears there.

中部的雕花：好像泛起的层层波纹。

Carving at the middle part: The carving is like the rolling waves.

垂花柱：腾龙纹雕刻栩栩如生。

Festoon pillar: The flying dragon patterns are carved vividly.

- **养心殿内的云龙纹镂雕垂花门**

The Openwork Festoon Door with Cloud and Dragon Pattern in the Hall of Mental Cultivation

江南园林及民居木雕

江南园林和民居建筑中的木雕是自然与艺术的完美结合。江南的建筑大都小巧而精致，因此建造者将精力注入到建筑的雕梁画栋上，木雕遍布建筑的每一个角落，且雕刻纤巧俊秀、古朴素雅、玲珑剔透，极具灵性。江南园林和民居建筑中的木雕题材主要有民间故事、戏曲人物、仙禽灵兽、花草鱼虫等，可谓十分丰富。

Wood Carving in the Gardens and Folk Houses in the Region South of the Yangtze River

The wood carving in the gardens and folk houses in the region south of the Yangtze River is a perfect combination of nature and art. Buildings in that reign are normally small and elegant. The builders thus shift their attention to the decoration, which includes carving and painting. The carvings are delicate, graceful, simple, elegant, dainty, and vivid. The wood carvings in the gardens and folk houses in the reign south of the Yangtze River involve the images of the figures from folk stories and operas, birds and beasts in fairytales, and flowers, grasses, fish, and insects of all kinds.

- **广州陈家祠门上的木雕**

  这是一种雅宅众宝吉祥图案，刻工细腻流畅，造型生动。

  The Wood Carving on the Door of the Chen's Ancestral Hall in Guangzhou

  This is an auspicious pattern featuring elegant house and numerous treasures. It features refined and smooth workmanship and vivid design.

- 山西王家大院门帘架上的“六合同春”镂空木雕

此木雕采用镂空手法雕刻出“六合同春”，即鹿、梅花、松树、喜鹊及花草图案，组成一幅春意盎然的图画，画面美观、刻工精致。

The Hollowed-out Wood Carving of Spring in the Air on the Door Curtain Frame of the Wang's Grand Courtyard in Shanxi Province

This wood carving is made with the hollowed-out technique. Entitled Spring in the Air, it contains patterns of deer, plum, pine tree, magpie, flower, and grass, forming a vivid picture of spring. It is indeed a beautiful artwork featuring the finest workmanship.

- 山西王家大院门帘架上的“三羊开泰”浮雕

在中文里，“阳”与“羊”同音。而在古代，“羊”即为“阳”。“三阳”即早阳、正阳、晚阳，均有勃勃生机之意。因此“三羊开泰”即寓意兴旺发达、万象更新。图案周围点缀有花草，使图案更加丰富。羊下巴上的胡子，使羊的形象更加生动。正中刻有太阳图案。松树寓意长寿。

The Relief of Three Sheep Opening up Good Luck on the Door Curtain Frame of the Wang's Grand Courtyard in Shanxi Province

In Chinese, sun and sheep share the same pronunciation *Yang*. In the ancient times, people regarded sheep as sun. Three Sun (Sheep) refers to the early sun, the highest sun, and the late sun, all indicating surging vitality. Three Sheep Opening up Good Luck stands for prosperity and renewal of all things. With decorative flowers and grasses around it, the pattern becomes richer. The beard on the chin of the sheep makes sheep image more vivid. In the middle is the pattern of the sun. The pine tree represents longevity.

- 五凤楼檐顶

五凤楼上的木雕，堪称徽州木雕的精华。

The Eaves Top of the Five-phoenix Building

The wood carving on the Five-phoenix Building is the essence of Huizhou wood carving.

- 屋檐上的木雕

此木雕以人物为表现对象，技术高超，人物动作变化多样，栩栩如生。

Wood Carving on the Eaves

This wood carving shows some vivid figures in diversified gestures.

• 徽州龙川胡氏宗祠五凤楼

徽州木雕向来以“多、精、美”著称，前额坊雕九个造型不同的狮子，称“九狮滚球遍地锦”。斗拱上雕有精美的云龙纹、牡丹纹等。

The Five-phoenix Building of the Hu's Ancestral Hall in Longchuan, Huizhou

Huizhou wood carving has always been famous for its large quantity, high quality, and great beauty. The forehead board has nine carved lions in different shapes, known as Nine Lions Rolling a Ball on Splendid Ground. The bucket arch is carved with exquisite patterns of cloud and dragon and peony.

• 山西王家大院门帘架上雕刻的十鹿图

在门帘有限的框架里，十头鹿的形象一一展现，鹿奔跑、跳跃的姿态刻画逼真，对十鹿不同方位的处理，使画面显得错落有致，疏密适宜。

The Ten-deer Painting Carved on the Door Curtain Frame of the Wang's Grand Courtyard in Shanxi Province

Within the limited frame of the door curtain, the images of ten deer are displayed vividly. Their running and leaping poses are vividly depicted. The handling of the deer postures from different angles results in a well-proportioned pattern.

尉迟恭：肤色黝黑，双目圆睁，姿态威武而严肃。

Yuchi Gong: He has a dark complexion. His two eyes are wide open.He is a powerful and serious general.

秦叔宝：凤眼白面，背扎四面旗，姿态安详而威严。

Qin Shubao: He has a pair of tilted eyes and a white face. His hand is combing his beard. Carrying four flags on his back, he shows a poised and serious look.

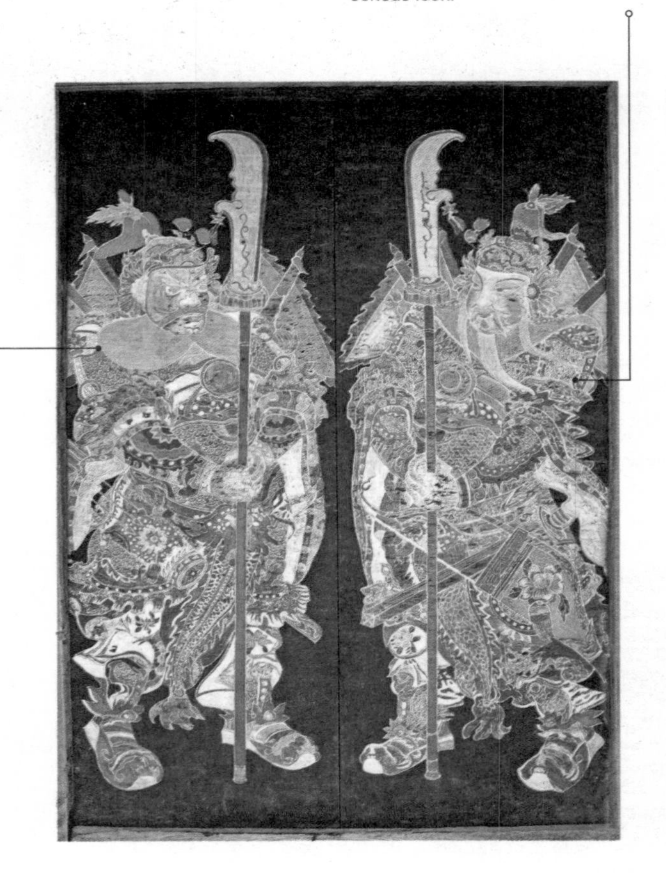

绘有门神的大门

大门上绘有秦叔宝、尉迟恭像。中国古代一般宅第大门的武将门神都是秦叔宝和尉迟恭，他们都是唐初著名的大将。

Doors Painted with Door-gods

The doors are painted with the portrait of Qin Shubao (also called Qin Qiong) and Yuchi Gong, common Door-gods in ancient China. They were renowned generals in the early Tang Dynasty.

> 家具雕刻

用于家具装饰的木雕，主要出现在八仙桌、太师椅、柜橱、梳妆台、坐墩、座屏、架子床、盆架以及婚嫁喜事的器物乃至祖宗牌位上。木雕图案包括狮、象、龙、凤以及各类吉祥图纹。这类木雕作品可以用来装点居室，古趣盎然，十分别致。也正是通过雕刻装饰，才使众多的古典家具产生了不同结构的造型美，拥有更大的魅力、时代感及民族特色。

明式家具木雕

明代家具追求简洁质朴、挺拔轻巧，因此以素面为主，利用木材本身的纹理来增加家具的美感。因此，明代家具较少雕饰，偶尔加以雕刻也是以线条为主，用宽窄、粗

> Furniture Carving

The wood carving for furniture decoration mainly appears on the eight-immortal table, master's chair, cabinet closet, dressing table, stool, pedestaled screen, canopy bed, basin stand, utensils used during the wedding, and even the memorial tablet of the ancestors. The wood carving patterns include that of lion, elephant, dragon, and phoenix, and other auspicious patterns. This type of wood carvings can be used to decorate houses. In fact, because of the presence of carving decoration, the classical furniture shows the modeling beauty of different structures and has greater glamour, stronger sense of their own era, and more distinct Chinese characteristics.

Ming-style Furniture Wood Carving

Ming-style furniture values simplicity, unaffectedness, forcefulness, and

细、长短、深浅不同的线来增加家具的线条变化，或用小面积的精致浮雕或镂雕以及镶嵌其他材质来增加装饰性，形成了明式家具木雕特有的风格。

lightness. Most of them are therefore not painted. The natural wood grains are left exposed to adding beauty to the furniture. Ming-style furniture does not have much carving decoration, either. The only carving, if any, is in form of lines. Either wide or narrow, thick or thin, long or short, dark or pale, these lines make the furniture full of changes. Sometimes, small area of delicate relief or fretwork, and inlay made of other materials are used to enrich the decoration effect and form the peculiar taste of the Ming-style furniture wood carving.

椅背：椅背简洁的镂空木雕，使整个圈椅显得古朴厚重。

Chair back: The chair back has simple and neat hollowed-out wood carving, which makes the entire chair unsophisticated and stately.

• 明代红木圈椅

A Mahogany Round-backed Armchair of the Ming Dynasty (1368-1644)

铁力木

铁力木材质坚硬、结构均匀、纹理致密，具有强度大、耐磨损、抗腐蚀、防虫蛀等特性，因此用途很广，是高级家具、特种雕刻、抗冲击器具、珍贵镶嵌和高级乐器的理想用材，比如造船、车辆、建筑等。铁力木家具做工讲究古朴天然，极少雕刻，通常只有简单的直线纹。因为铁力木纤维较粗，性韧，易皲裂，且材质非常硬，所以修光打磨都非常不容易，工匠在制作铁力木家具时，会充分利用铁力木古拙淳朴的特质进行加工，将纹饰留粗，这是其他木制家具中不常见的。

Ironwood

Ironwood has a hard texture, an even structure, and compact grains. It is strong and performs good in resisting wear, corrosion, and worm damage. Therefore, it is widely used as an ideal material for making first-class furniture, special carving, impact-proof utensils, precious inlay, and first-class musical instrument. It is also used to make boat and cart, also as building structures. Ironwood furniture bears little carving, but only a few simple straight lines. The wood fiber is thick, tenacious, hard, and easy to crack, making grinding and polishing a difficult job. When making ironwood furniture, the carpenters will fully use the wood's simple and pure properties and leave only thick line decoration. This is a feature uncommon among furniture made of other wood.

铁力木的切面

A Section of Ironwood

明代铁力木罗汉床

工匠在制作铁力木家具时，充分利用铁力木的木性特征进行加工，雕刻线条较粗，彰显出铁力木古拙淳朴的特质。因为其木纤维较粗长，性韧，不易切断，所以雕刻时易起毛刺，且材质硬，不易打磨。床围上攒接着错落有致的几何图案，虚实相间。

A Ironwood Luohan Bed of the Ming Dynasty (1368-1644)

When making ironwood furniture, the carpenters make full use of the wood's simple and pure properties by carving relatively thick lines. Ironwood has thick and long wood fibers, which are too tenacious to cut and too hard to grind. As a result, burr is often seen during carving. On the railing of this bed, there are well-proportioned geometric patterns.

清式家具木雕

清代家具追求繁缛的风格，造型庄重、体量庞大、雕饰繁琐。这一时期的家具木雕艺术发展空前，技艺精良，以雕刻、镶嵌、描金等为主，大多采用多种工艺结合的方法。同时，清代受欧洲等国艺术风格的影响，木雕装饰更加华丽，雕琢更加细腻，形式更加夸张，使家具装饰达到雍容华贵的效果，形成了清式家具木雕特有的风格。

Qing-style Furniture Wood Carving

Furniture makers in the Qing Dynasty (1616-1911) highly valued sophisticated style, stately design, large size, and detailed carving decoration. Furniture wood carving developed to an unprecedented height during this period. Qing-style furniture showed refined techniques such as carving, inlaying, and gold tracing, which were often used in combination. Meanwhile, influenced by the artistic styles of European countries, the wood carving decoration in the Qing Dynasty became more gorgeous and exaggerated and formed its unique style.

中国传统木雕图案的寓意

在中国传统木雕图案中，每种物品都有它固定的寓意，比如牡丹寓意富贵，石榴寓意多子，鹿寓意福禄，蝙蝠、桃子和佛手寓意多福，喜鹊寓意喜庆，鱼寓意富足，瓶寓意平安，橘子寓意吉庆，大象寓意吉祥，葫芦寓意子孙万代绵长等。还有一些常见的复杂图案，如“百事如意”，即由柏、柿子、如意或灵芝组成，喻事事称心如意。再如“岁寒三友”，即以松、竹、梅三种不畏严寒的植物构成图案，多为古代文人所用，喻坚韧不拔，也象征友谊长存。

Meaning of the Traditional Chinese Wood Carving Patterns

In traditional Chinese wood carving patterns, every article has its fixed meaning. For example, peony stands for wealth, pomegranate for many sons, deer for luck and wealth, bat, peach, and Buddha hand for many lucks, magpie for happiness, fish for richness and abundance, bottle for peace and safety, orange for auspice, elephant for luck, and calabash for long-lasting posterity. There are also some commonly seen complicated patterns. For example, the one entitled All Things Going As Wished is composed of cypress, persimmon, ruyi or ganoderma. Another pattern, Three Cold-weather Friends, is composed of three plants thriving in winter, namely pine, bamboo, and plum. The ancient scholars often used the pattern to imply tenacity and long-term friendship.

柜门上的“四福（蝠）捧寿”纹：左右门扇的上面各浮雕“四福（蝠）捧寿”纹，四只蝙蝠环绕团寿飞翔。“蝠”与“福”谐音，寓意福气、幸福。

The pattern of Four Lucks Carrying Longevity on the cabinet doors: The relief pattern of Four Lucks Carrying Longevity is on both doors. Four bats fly around the round Chinese character longevity. Bat and luck sound the same in Chinese.

柜门上的“幸福（蝠）吉祥”纹：左右门扇的下面各浮雕“幸福（蝠）吉祥”纹，正在飞翔的蝙蝠寓意福在眼前，幸福吉祥。

The pattern of Luck and Happiness on the cabinet doors: The relief pattern of Luck and Happiness is on the lower parts of both doors. The flying bats indicate that luck and happiness are at hand.

清代红木雕花万历柜

A Mahogany Carved Wanli Display Cabinet of the Qing Dynasty (1616-1911)

清代紫檀木浮雕屏风

A Red Sandalwood Relief Screen of the Qing Dynasty (1616-1911)

桌围的镂空雕刻：桌围采用镂空雕刻，突出“福寿三多”图案中的佛手、桃和石榴。

Hollowed-out carving on table railing: The table railing is made with hollowed-out carving to accentuate the Buddha hand, peach, and pomegranate in the pattern of Three Items of Great Luck and Longevity.

案腿的浮雕：使“福寿三多”图案和如意纹显得很有立体感。

Relief on the table leg: The relief gives a 3D effect to the pattern of Three Items of Great Luck and Longevity and the pattern of ruyi, an S-shaped ornamental object for good luck.

清代红木条案上的木雕

Wood Carving on a Mahogany Long Table of the Qing Dynasty (1616-1911)